Emerging Challenges and Opportunities of High Speed Rail Development on Business and Society

Raj Selladurai
Indiana University Northwest, USA

Peggy Daniels Lee
Kelley School of Business, Indiana University, USA

George VandeWerken
Providence Bank, USA

A volume in the Advances in Civil and Industrial Engineering (ACIE) Book Series

ENGINEERING SCIENCE REFERENCE
An Imprint of IGI Global

Published in the United States of America by
Engineering Science Reference (an imprint of IGI Global)
701 E. Chocolate Avenue
Hershey PA 17033
Tel: 717-533-8845
Fax: 717-533-8661
E-mail: cust@igi-global.com
Web site: http://www.igi-global.com

Library of Congress Cataloging-in-Publication Data

Names: Selladurai, Raj, 1954- editor. | Lee, Peggy Daniels, 1948- editor. | VandeWerken, George, 1947-
Title: Emerging challenges and opportunities of high speed rail development on business and society / Raj Selladurai, Peggy Daniels Lee, and George VandeWerken, editors.
Description: Hershey : Engineering Science Reference, 2016. | Includes bibliographical references and index.
Identifiers: LCCN 2016002412| ISBN 9781522501022 (hardcover) | ISBN 9781522501039 (ebook)
Subjects: LCSH: High speed trains--United States. | High speed ground transportation--Case studies.
Classification: LCC HE2741 .E54 2016 | DDC 303.48/32--dc23 LC record available at http://lccn.loc.gov/2016002412

This book is published in the IGI Global book series Advances in Civil and Industrial Engineering (ACIE) (ISSN: 2326-6139; eISSN: 2326-6155)

British Cataloguing in Publication Data
A Cataloguing in Publication record for this book is available from the British Library.

Advances in Civil and Industrial Engineering (ACIE) Book Series

ISSN: 2326-6139
EISSN: 2326-6155

Mission

Private and public sector infrastructures begin to age, or require change in the face of developing technologies, the fields of civil and industrial engineering have become increasingly important as a method to mitigate and manage these changes. As governments and the public at large begin to grapple with climate change and growing populations, civil engineering has become more interdisciplinary and the need for publications that discuss the rapid changes and advancements in the field have become more in-demand. Additionally, private corporations and companies are facing similar changes and challenges, with the pressure for new and innovative methods being placed on those involved in industrial engineering.

The **Advances in Civil and Industrial Engineering (ACIE) Book Series** aims to present research and methodology that will provide solutions and discussions to meet such needs. The latest methodologies, applications, tools, and analysis will be published through the books included in **ACIE** in order to keep the available research in civil and industrial engineering as current and timely as possible.

Coverage

- Transportation Engineering
- Productivity
- Quality Engineering
- Coastal Engineering
- Materials Management
- Ergonomics
- Operations Research
- Optimization Techniques
- Hydraulic Engineering
- Engineering Economics

IGI Global is currently accepting manuscripts for publication within this series. To submit a proposal for a volume in this series, please contact our Acquisition Editors at Acquisitions@igi-global.com or visit: http://www.igi-global.com/publish/.

The Advances in Civil and Industrial Engineering (ACIE) Book Series (ISSN 2326-6139) is published by IGI Global, 701 E. Chocolate Avenue, Hershey, PA 17033-1240, USA, www.igi-global.com. This series is composed of titles available for purchase individually; each title is edited to be contextually exclusive from any other title within the series. For pricing and ordering information please visit http://www.igi-global.com/book-series/advances-civil-industrial-engineering/73673. Postmaster: Send all address changes to above address.

Titles in this Series

For a list of additional titles in this series, please visit: www.igi-global.com

Handbook of Research on Applied E-Learning in Engineering and Architecture Education
David Fonseca (La Salle Campus Barcelona, Universitat Ramon Llull, Spain) and Ernest Redondo (Universitat Politècnica de Catalunya, BarcelonaTech, Spain)
Engineering Science Reference • copyright 2016 • 569pp • H/C (ISBN: 9781466688032) • US $310.00 (our price)

Emerging Design Solutions in Structural Health Monitoring Systems
Diego Alexander Tibaduiza Burgos (Universidad Santo Tomás, Colombia) Luis Eduardo Mujica (Universitat Politecnica de Catalunya, Spain) and Jose Rodellar (Universitat Politecnica de Catalunya, Spain)
Engineering Science Reference • copyright 2015 • 337pp • H/C (ISBN: 9781466684904) • US $235.00 (our price)

Robotics, Automation, and Control in Industrial and Service Settings
Zongwei Luo (South University of Science and Technology of China, China)
Engineering Science Reference • copyright 2015 • 337pp • H/C (ISBN: 9781466686939) • US $215.00 (our price)

Using Decision Support Systems for Transportation Planning Efficiency
Ebru V. Ocalir-Akunal (Gazi University, Turkey)
Engineering Science Reference • copyright 2016 • 475pp • H/C (ISBN: 9781466686489) • US $215.00 (our price)

Contemporary Ethical Issues in Engineering
Satya Sundar Sethy (Indian Institute of Technology Madras, India)
Engineering Science Reference • copyright 2015 • 343pp • H/C (ISBN: 9781466681309) • US $215.00 (our price)

Emerging Issues, Challenges, and Opportunities in Urban E-Planning
Carlos Nunes Silva (University of Lisbon, Portugal)
Engineering Science Reference • copyright 2015 • 380pp • H/C (ISBN: 9781466681507) • US $205.00 (our price)

Technology and Practice in Geotechnical Engineering
Joseph B. Adeyeri (Federal University Oye-Ekiti, Nigeria)
Engineering Science Reference • copyright 2015 • 835pp • H/C (ISBN: 9781466665057) • US $225.00 (our price)

Editorial Advisory Board

List of Reviewers

Table of Contents

Section 3

Detailed Table of Contents

California has started construction on the first true high-speed rail system in the United States. In this chapter, the CEO of the California High-Speed Rail Authority describes how the state's historic investment in new high-speed rail infrastructure will benefit Californians. Benefits include job creation, statewide modernization of existing local rail lines, short and long-term reductions in greenhouse gas emissions, preservation of agricultural land and environmentally sensitive habitats, advanced clean and green construction practices and technologies, a commitment to 100% renewable energy use, and assistance to partner cities on transit-oriented planning for high-speed rail station areas.

For over 50 years, after both public and private studies confirmed California's need for high-speed rail, nothing happened. The rest of the US had a similarly disappointing experience when attempting to break into the rapidly growing family of nations with operating high-speed trains. Yet, over the past year, California has contracted over $2.5 billion in high-speed rail civil work, is preparing RFPs for three times that amount, and has secured commitments for nearly $25 billion in additional federal and state funding over the next 30 years. Private investors are lining up for the nation's first high-speed rail public-private partnership to complete this $68 billion, 540 mile project. This chapter describes the unprecedented technical excellence, political courage, and visionary voter support that created that high-speed rail breakthrough for California and the nation.

Section 2

Transportation planners in the Midwest are advancing an agenda for high-speed rail while simultaneously investing in numerous corridors that have different strengths and weaknesses. To assist in establishing priorities, this paper compares the size of the travel markets in seven of these Midwestern corridors. Using a technique known as the gravity model, it demonstrates that the corridors linking Chicago to Cincinnati and Detroit likely exceed the others on the basis of total passenger traffic, and, along with the Chicago – Twin Cities route, rank at the top with respect to the passenger-miles of travel. The routes to Cleveland, St. Louis, and other locales compare less favorably. The top corridors, while shown to be only a fraction of the size of the California and Northeast Corridor routes, are found to be appreciably larger than most of the country's proposed high speed rail corridors, suggesting they have considerable potential for rail-passenger development.

By conservative, reliable, objective, and preliminary estimates, the United States could generate at least $660 million annually – $26.4 billion over four decades – by building and operating faster, more frequent, more dependable, and more highly integrated passenger rail service in four corridors: The Northeast; the Chicago Hub; California; and the Northwest. Numerous studies by a variety of national and international organizations and government agencies suggest that the actual cost of building and/or improving passenger rail service lines is significantly less than the cost per mile of other modal alternatives. In fact, in many corridors, passenger rail is the only feasible option for adding capacity, given the practical constraints facing aviation and highway expansion. Further, by integrating improved, higher speed intercity passenger rail service into the existing transportation system, major issues of congestion, mobility and economic inequality can be addressed.

As we witness the advancement of U.S. high-speed rail initiatives, the country can look towards its European and Asian counterparts for best practices and lessons learned from their decades of high-speed rail design and operations. These experiences gained may be applicable towards projects such as the Texas Central Railway and the California High-Speed Rail Project. This chapter will address the events of 2009 that have brought domestic high-speed rail to the forefront of U.S. rail transportation. This includes the new FRA Tier I and proposed Tier III criteria, challenges associated with each FRA tier of operation, overseas interoperability efforts, snapshots of international experiences (from policy and technological perspectives), the holistic system-based approach to safety, ongoing efforts of the FRA Engineering Task Force, and additional challenges and opportunities moving forward.

This chapter examines safety related to high and higher speed inter-city passenger rail operations. Generally, most risk related to rail travel is associated with collisions between on-track equipment and vehicles and pedestrians at level crossings. A larger amount of fatalities, but not necessarily systemic risk, is due to trespassing on the rail right-of-way. A third aspect of risk related to rail operations lies in rail to rail collisions and derailments, some of which may be catastrophic. A discussion of safety programs implemented by the rail industry at national and international levels concludes the chapter.

Endorsements

This compendium of essays/discussions on High Speed Rail development addresses a quickly emerging area of mass transit that has been successful in other parts of the world. The articles come from an outstanding mix of highly placed/respected practitioners, scientists, and consultants. The very first chapter has been written by the world famous CEO of the California High Speed Rail Authority, Jeff Morales. He provides us insights based on his experience at the highest levels of high-speed rail (HSR) management. Other early chapters include an historical overview of development of HSR in California, which has languished for years, but has suddenly accelerated by a major governmental investment of time and resources. Noted scientists from DePaul University, Indiana University and University of Calgary contribute significant analyses of other transportation corridors and their suitability for HSR development, the relationship of HSR to the airline industry, and the status of high speed rail development in Canada, as well as future implications of HSR development. The issues of system safety, the cost of NOT developing HSR, and future implications for the industry are discussed in other critical essays.

Cuthbert L. Scott III, Ph.D.
Associate Dean & Associate Professor Emeritus, Indiana University Northwest, USA

This book is a must-read for anyone interested in understanding the complex dynamics of high-speed rail (HSR) systems. The editors have done a fine job in soliciting chapters from the world's foremost experts on HSR. In-depth case studies from Europe, Asia, Canada, and the US provide invaluable data about the essential determinants of success. Dr. Selladurai and his colleagues have offered a compelling summary of the available research on HSR and insightful suggestions about the future potential of this amazing engine of economic growth.

Charlie Hobson, Ph.D.
Professor, School of Business and Economics, Indiana University Northwest, USA

Foreword

As Chairman of the California High-Speed Rail Authority, one of the questions that I'm often asked is: "How can we afford to build high-speed rail?" While I understand concerns about government budgets and smart spending, I often respond to such questions with a counter-question: "How much will we spend *not* to build high-speed rail?"

Indeed, here in California (where we already boast the largest population in the United States) our populace is expected to grow to some 50 million by the year 2050. To put that number in context, we'll effectively be adding the entire population of New York over the next 35 years. That would all be well and good if our transportation infrastructure was serving our residents adequately and if we had plans to ensure that it does so in the future. Unfortunately, the opposite is the case. California's 170,000 miles of roadway are the busiest in the nation, with estimates pegging the cost of time lost and fuel wasted in traffic congestion at $18.7 billion annually. Moreover, the Los Angeles to San Francisco air market is the busiest short-haul market in the nation with over 5 million passengers annually. This had led to approximately one out of every four flights being delayed by an hour or more.

These congestion problems aren't new, and they certainly aren't going away anytime soon as our population continues to grow. So why don't we expand our highways and airports to alleviate these issues instead of implementing a high-speed rail system? Because achieving the capacity that high-speed rail offers through new freeway miles and airport runways would cost significantly more – up to 3 times more - continue to contribute to harmful greenhouse gas emissions, and ultimately prove unsustainable. When looked at in this light, it becomes clear that the transportation problems we face here and California – and throughout the nation – cannot be solved by a "more of the same" approach.

Aside from the realities of our over-taxed infrastructure and growing population, high-speed rail fills a gap in our transportation network that has existed for decades: medium-distance trips. Specifically, for trips between 100 and 600 miles, automobile and air travel has become inefficient when measured in cost, time, energy, and emissions. These are trips that, for the average traveler, are just too long to drive but not quite short enough to go through the expense and process of air travel. International experience has shown that when major cities separated by comparable distances, like Los Angeles and San Francisco, are connected via high-speed rail the majority of travelers will switch modes from air or car to rail. For example, the Madrid to Seville line in Spain has increased rail travel between those cities from 16% to 51% of all trips, while total traffic between those cities has increased by 35%; this figure indicating that high-speed rail has actually encouraged more trips between these metropolitan hubs.

But these conditions that make high-speed rail a crucial investment in California's future exist throughout the nation. There are "megaregions" across the country that face the same challenges that California faces and possess similar attributes and strengths. These are areas with large or dense populations that boast significant economic capacity where highly skilled talent, scientific achievement, and technological innovations are concentrated and compete on a global scale. From the Northeast to the Northwest, the Great Lakes to the Gulf Coast, and from Florida to Texas, these regions produce billions – and sometimes trillions – of dollars in economic output, but can be stifled by lack of efficient connectivity. It will be absolutely critical to America's continued economic growth that we foster the effective transfer and interaction of people, materials, and ideas to ensure that we optimize efficiencies within and between our megaregions. High-speed rail offers a means by which to achieve these goals.

This is why it's so exciting to see high-speed rail gaining momentum throughout the nation and to participate in efforts like this one: *Emerging Challenges and Opportunities of High Speed Rail Development on Business and Society*. By continuing to expound on the many challenges and benefits of high-speed rail and advocating for its expansion, efforts such as this one can have a real, substantive impact on public policy debates that move high-speed rail forward. We're already seeing exciting developments in Florida, Texas and Nevada. But only with continued efforts like this one will we see a truly nationwide high-speed rail network in America.

Here in California, construction is now underway on more than 100 miles of brand new high-speed rail infrastructure in our Central Valley, with more than $2 billion in construction contracts already executed. Right now, the first vertical structure associated with the project - the Fresno River Viaduct over the Madera River - is going up just north of Fresno. In the coming months, residents and visitors to the Central Valley will start to see more structures and activity along the backbone of the system between Madera and Bakersfield

At the same time, we are moving aggressively on planning and implementing high-speed service in Northern and Southern California. In the Bay Area we are working closely with our partners at the intercity train operator Caltrain to implement a blended system between San Jose and San Francisco. This blended system will benefit commuters and regional travelers in the short-term by electrifying the corridor and implementing Positive Train Control, while paving the way for high-speed rail service primarily on shared track in the future as our system extends to the San Francisco Peninsula.

We are also working with transportation agencies in the Los Angeles Basin, including LA Metro and Metrolink, to upgrade existing systems and improve connections to begin the process of implementing high-speed rail. Both of these efforts are part of our phased approach to reaching a statewide high-speed rail system, wherein individual investments are made on a piece by piece basis and each have independent utility to commuters and travelers throughout the state. This method is similar to the one used to implement the interstate highway system, and we believe it can be used as a model for the rest of the nation as other states begin to move toward high-speed rail. Our approach takes into account the realities of government budgets and spending and provides value at each step along the way to achieving state-of-the-art rail systems.

It is also important to add that the development of high speed rail in California is a key tool of our state's program to contribute to global reduction of greenhouse gas emissions. Governor Jerry Brown has been an ardent champion of the need for governments at all levels to pursue policies that combat climate change. California's initiative to develop a modern, electric rail system, including high speed intercity as well as regional rail, combined with good land use decisions around the stations, will have a material impact on greenhouse gas emissions.

In closing, I would like to thank Dr. Raj Selladurai for inviting me to pen a modest foreword to this important academic work. As I have explained, the need for high-speed rail is clear and we know that it can provide a sustainable solution to some of our transportation problems. The key now is to continue to explain, educate, and publicize this information to the broadest spectrum possible. I strongly believe that there is a great deal of momentum behind high-speed rail in the United States and that the trend will only continue with the continued advocacy this book represents. Someday we will have a modern, efficient high-speed rail network in America. We can't afford not to.

Dan Richard
California High-Speed Rail Authority, USA

Dan Richard *is the Chair of the Board of Directors for the California High-Speed Rail Authority, USA.*

Preface

High-speed rail is clearly very popular and extensively prevalent in several other countries from Europe and Asia especially China, which has exhibited unprecedented growth in the short time since its high-speed rail inception. Similarly, in the U.S. too, high-speed rail will provide a cost-effective alternative to highway and airline transportation bringing with it significant savings of travel time, improved travel experience, and enhanced traveler productivity in many travel corridors in the United States. The pressures for such a solution continue to mount due to numerous emerging factors such as:

- Growing population density and its impact on transportation infrastructure
- Increasing congestion on highways and at airports
- Increased fuel cost impact on highway and airline travel
- Mounting costs for highway construction and maintenance
- Decreasing fuel tax revenues to fund highway trust funds and political stalemate to address the issue
- Limited availability of or exorbitant cost to acquire more property for highway expansion
- Increased significance of weather-related disruptions of existing transportation modes
- Increasing costs for highway tolls and downtown parking fees
- Growing demand for improved travel access directly to urban core business districts
- Increasing burden of non-productive work time spent in highway and airport congestion
- Need to develop more environmental-friendly transportation alternatives.

All these point to the importance of considering the viability of high-speed rail as a beneficial social, environmental, and economic alternative to highway and air travel. It may be time for all stakeholders in California, Indiana/Midwest as well as the rest of the USA to consider seriously other alternatives to supplement travel by road and air. Based on a recent state/regional focus group study's findings (Selladurai, 2013) as well as several other studies, there is clearly a need for a research publication like this, which documents the case for high-speed rail as a viable option to supplement road travel in California, Indiana, Midwest, and the US. This book, "Emerging Challenges and Opportunities of High Speed Rail on Business and Society," is certainly a comprehensive reference resource appropriate for such a time as this, and we believe that it would make a major contribution toward facilitating/strengthening the promotion, support, and implementation of the high-speed rail vision into reality. Recent, current, ongoing and future planned high-speed rail related developments especially in California, Florida, and Minnesota and other rail corridors provide us a wonderful snapshot of future things to come! High Speed Rail would certainly be the "LINK" to the future!

The purposes of "Emerging Challenges and Opportunities of High Speed Rail on Business and Society," are to:

- Provide a forum for discussing and sharing high quality research and analyses by scholars and practitioners from all over the world on topics related to high-speed rail.
- Present a variety of basic and practical research studies for stimulating further research related to high-speed rail.
- Integrate and synthesize interdisciplinary, theoretical, and practical research studies related to high-speed rail.
- Formulate and present conceptual models for research on high-speed rail and situations.
- Serve as a catalyst for policy makers, decision-makers, and high-speed rail stakeholders including academicians, developers, business and organizational leaders, managers, professionals, and employees in public and private organizations to help formulate and implement plans for high-speed rail as a viable transportation option in the U.S.

Policy makers, academicians, researchers, advanced-level students, technology developers, private companies and investors, and government officials at federal, state, and local levels will find this text useful in furthering their research exposure to pertinent topics in high speed rail transportation and assisting in furthering their own research efforts in this field. Also, this publication would educate and influence business, corporate, and other organizational leaders, managers, and employees of various types of organizations about the need, impact, formulation, and implementation plans for high-speed rail in transportation.

"Emerging Challenges and Opportunities of High Speed Rail on Business and Society," makes a significant contribution to the growing vision of high-speed rail. It brings together interesting perspectives from leaders, scholars, and experts about high-speed rail in various countries of the world including Canada, Europe, Asia, and the U.S. Although several chapters are of scholarly nature coming from academicians, the topics discussed are of keen interest and high value to a variety of leaders, managers, and practitioners from all types of organizations. It integrates top quality research on high-speed rail from variety of situations and settings into a single source.

The book provides a wide array of interesting and timely topics in the area of high-speed rail, and it would be a valuable primary reference source for graduate students and undergraduate students interested in high-speed rail and transportation.

"Emerging Challenges and Opportunities of High Speed Rail on Business and Society," focuses on three main areas of high-speed rail. The first section focuses on current developments and implementation plans in high-speed rail especially in California. The second section looks at various issues, challenges, and opportunities in high-speed rail development. These include the market size and potential of high-speed rail corridors in the Midwest; the tremendous benefits of high-speed rail and the huge negative implications of not developing high-speed rail as found extensively in most of the rest of the world; and the operational issues related to high-speed rail. Further, it includes the safety issues and concerns; the airlines-high-speed rail collaboration model; the blended approach model; and the potential of high-speed rail in Canada. The third section provides for us an overview of the past, present, and future implications of high-speed rail in the United States.

In the opening chapter, "California High-Speed Rail: A Transformative Investment in California's Future," Jeff Morales focuses on the status of high-speed rail in California. He discusses how California has started construction on the first true high-speed rail system in the United States. The CEO of the California High-Speed Rail Authority describes how the state's historic investment in new high-speed rail infrastructure will benefit Californians. He discusses the many benefits that include job creation, statewide modernization of existing local rail lines, short and long-term reductions in greenhouse gas emissions, preservation of agricultural land and environmentally sensitive habitats, advanced clean and green construction practices and technologies, a commitment to 100% renewable energy use, and assistance to partner cities on transit-oriented planning for high-speed rail station areas.

Continuing in similar fashion, in the next chapter "California's "Fast-Track" to High-Speed Rail: The Early Challenges and Ultimate Success of the California High-Speed Rail Project," California High-Speed Rail Authority leaders and transportation experts, Rod Diridon and Ben Tripousis, highlight the development of high-speed rail development over the past several years. For over 50 years, after both public and private studies confirmed California's need for high-speed rail, nothing happened. The rest of the US had a similarly disappointing experience when attempting to break into the rapidly growing family of nations with operating high-speed trains. Yet, over the past year, California has contracted over $2.5 billion in high-speed rail civil work, is preparing RFPs for three times that amount, and has secured commitments for nearly $25 billion in additional federal and state funding over the next 30 years. Private investors are lining up for the nation's first high-speed rail public-private partnership to complete this $68 billion, 540 mile project. This chapter describes the unprecedented technical excellence, political courage, and visionary voter support that created that high-speed rail breakthrough for California and the nation.

In the next chapter, "Comparing the Size of the Market in Potential High Speed Rail Corridors of the Midwest: A Gravity Model Approach," Professor Joseph Schwieterman evaluates the size and potential of several travel markets in the Midwest. He contends that transportation planners in the Midwest are advancing an agenda for high-speed rail while simultaneously investing in numerous corridors that have different strengths and weaknesses. To assist in establishing priorities, his chapter compares the size of the travel markets in seven of these Midwestern corridors. Using a technique known as the gravity model, it demonstrates that the corridors linking Chicago to Cincinnati and Detroit likely exceed the others based on total passenger traffic, and, along with the Chicago-Twin Cities route, rank at the top with respect to the passenger-miles of travel. The routes to Cleveland, St. Louis, and other locales compare less favorably. The top corridors, while shown to be only a fraction of the size of the California and Northeast Corridor routes, are found to be appreciably larger than most of the country's proposed high-speed rail corridors, suggesting they have considerable potential for rail-passenger development.

Eric Peterson, a transportation policy consultant, in this subsequent chapter, "Benefits Forgone: The Cost of Not Building Higher Performing Passenger Rail" emphasizes the tremendous benefits of high-speed rail. He points out that by conservative, reliable, objective, and preliminary estimates, the United States could generate at least $660 million annually – $26.4 billion over four decades – by building and operating faster, more frequent, more dependable, and more highly integrated passenger rail service in four corridors: The Northeast; the Chicago Hub; California; and the Northwest. Numerous studies by a variety of national and international organizations and government agencies suggest that the actual cost of building and/or improving passenger rail service lines is significantly less than the cost per mile of other modal alternatives. In fact, in many corridors, passenger rail is the only feasible option for adding capacity, given the practical constraints facing aviation and highway expansion. Further, by integrating improved, higher speed intercity passenger rail service into the existing transportation system, major issues of congestion, mobility and economic inequality can be addressed.

Further, transportation consultant experts Francis Banko and Jackson Xue discuss some other significant challenges and opportunities for high-speed rail development in their chapter, "Emerging Challenges and Opportunities for U.S. High-Speed Rail Development." They recommend that as we explore various U.S. high-speed rail initiatives, the country can look towards its European and Asian counterparts for best practices and lessons learned from their decades of high-speed rail design and operations. These experiences gained may be applicable towards projects such as the Texas Central Railway (with envisioned revenue service start of 2020) and the California High-Speed Rail Project (with envisioned revenue service start of 2022). In their chapter, they address the events of 2009 that have brought domestic high-speed rail to the forefront of U.S. rail transportation. This includes the new FRA Tier I and proposed Tier III criteria, challenges associated with each FRA tier of operation, overseas interoperability efforts, snapshots of international experiences (from policy and technological perspectives), the holistic system-based approach to safety, ongoing efforts of the FRA Engineering Task Force, and additional challenges and opportunities moving forward.

The next chapter focuses on another important, timely topic namely security. In his chapter, "Safety of Domestic High Speed Passenger Rail Operations: Safety is Good Business," Steve Laffey examines safety related issues to high and higher speed inter-city passenger rail operations. Generally, most risk related to rail travel is associated with collisions between on-track equipment and vehicles and pedestrians at level crossings. A larger amount of fatalities, but not necessarily systemic risk, is due to trespassing on the rail right-of-way. A third aspect of risk related to rail operations lies in rail-to-rail collisions and derailments, some of which may be catastrophic. A discussion of safety programs implemented by the rail industry at national and international levels concludes the chapter.

In continuing to look at opportunities and challenges, authors Peggy Lee, George VandeWerken, and Raj Selladurai have developed an airline-high-speed rail collaboration model in their chapter, "Exploring the Airline-High Speed Rail Collaboration Model: Efficient Service and Mutual Benefits." They have shown that short-haul airline passenger traffic (less than 500 miles) is decreasing nationwide. This decline may be attributed to legacy airlines' rising costs (especially fuel), increased airport congestion, and increased travel time due to post-September 2001 TSA security screening. Previous studies tend to look at the substitution of high-speed rail and other transportation modes for air travel, especially as travel times shorten. Substitution usually takes the form of collaboration or competition between competing modes and competing carriers. The authors in this chapter present an alternative view – with a discussion of the proposition that air carriers may benefit more from collaborative arrangements that allow them to "own" at least a portion of the intermodal passenger experience rather than shifting or transferring passengers to competitive non-air modes. They believe that this proposal has merit and is worth consideration, and then the chapter concludes with a research agenda designed to test the proposition empirically.

Further, Rick Harnish, the Midwest High Speed Rail Association director, explores the significant issue of track structure use for rail systems in the U.S. In his chapter, "Blended Approach: Efficient Track Structure for High Speed Rail," the author contends that a common misconception exists that high-speed rail is a stand-alone system, separate from the conventional network. In fact, only Taiwan's system is completely separate. All other systems use the "blended approach" in which trains use a combination of high-speed and conventional tracks in a single journey. It is similar to cars and buses using both Interstate Highways and local roads in a single trip. Each nation's railroad has taken a different approach to developing its high-speed networks in response to diverse needs. The common thread is that by implementing individual projects as a series of building blocks towards a long-term plan, immediate improvements can be enjoyed while working towards a very powerful system. He then looks at various countries' track system including Japan, France, Germany, England, Spain, and others as well as makes some recommendations for high-speed rail in the U.S. He concludes that for high-speed rail to develop in the United States, a comprehensive blended approach will be needed, with initial focus placed on a few key segments.

Next, we move outside our boundaries to explore the potential of high-speed rail for our nearest northern neighbor, the country of Canada. Professors Jaydeep Balakrishnan, Tianyuan Zhu, and Peter Wallis in their chapter, "High Speed Rail in Canada: Feasibility Studies, Influential Factors and International Comparisons," state that High Speed Rail (HSR) is gaining popularity around the world for a variety of reasons. However, North America lags behind and in particular Canada, where there is no HSR. There has been discussion of implementing HSR in Canada since the 1970s in two possible corridors (one in central Canada and the other in Alberta). In this chapter, the feasibility studies that have been done on both corridors are discussed. In this respect the focus is on the recent studies, from different sources including the Alberta Legislature, Ontario and Quebec Governments, consulting companies, and a non-profit transportation institute. The factors that influence the viability of HSR are discussed and reference is made to HSR implementations in other countries that could inform the development of HSR in Canada. On a positive note, it appears that the Ontario government plans to move ahead with HSR in part of the central Canadian corridor.

In the third, final section of the book, authors Raj Selladurai and George VandeWerken begin by exploring the current and future state of high-speed rail. In their chapter, "High Speed Rail: Study, Report, and Future Considerations" they discuss the recent Southwest states' multi-state planning collaborative study, which the Federal Railroad Administration initiated, coordinated, and released in 2015. It highlights the process, planning context, conclusions, recommendations, and implications, which may serve as a reference model for future studies in other regions of the United States. The chapter then looks at the momentum of high-speed rail in the U.S especially in California and explores some of the implications of high-speed rail for the nation in the future.

Then, Dorothy Hagan, City of Fort Wayne, Indiana summarizes the status of high-speed rail in Northeast Indiana. In her chapter, "Progress through Partnerships: Northeast Indiana Passenger Rail Association," she provides a brief history of the development of the Northeast Indiana Passenger Rail Association (NIPRA). She presents an overview of two studies utilized by NIPRA to develop its preferred rail corridor through Fort Wayne. She discusses the partnerships that were developed at the local level across the corridor; the study completed to provide a business case for the corridor; discussions on coordination and fundraising efforts by NIPRA and the City of Fort Wayne staff, including a Federal Railroad Administration (FRA) grant application; and the next steps for NIPRA and corridor partners.

In the next chapter, Dennis Hodges, a pioneer of high-speed rail with the Indiana Passenger Rail Alliance, looks at the vision and potential reality of high-speed rail. This chapter, "High Speed Rail: Are we there yet?" addresses a vision of what was, is, and can be. Japan is already 65 years into its passenger rail improvements and modifications. France is just over 35 years and China is eight (8) years into its passenger rail improvements and modifications. Why cannot the United States see the economic and environmental reasons for doing the same in this country? This chapter discusses that and more. The author has attempted to parcel out the whys, what's and where's as to how the United States became so parochial when it comes to passenger rail transportation; why it is so reticent towards moving forward, and what can be done to change the current political and social attitude and climate.

Finally, the authors conclude the book with the chapter, "High Speed Rail: Suggestions for Case Studies, Research Questions, and Topics. Here the authors Raj Selladurai and George VandeWerken focus on some suggestions for case studies, research questions, and topics related to high-speed rail especially for use in the academic and professional settings. They recommend that the students, professionals, policy makers, and all high-speed rail stakeholders would benefit from these resources, which would facilitate them to make significant contributions to the development of high-speed rail in the U.S. Further, a focus on these would help stimulate more exploring, thinking, discussing, and researching into innovative ideas, plans, and strategies toward the formulation, implementation, and evaluation of high-speed rail for the U.S. for several years in the future.

We believe that our book, "Emerging Challenges and Opportunities for U.S. High-Speed Rail Development" is a new (one of very few), unique, significant resource currently available in the market. It would be a useful resource toward developing policies, decision-making, formulation, and implementation of high-speed rail in the U.S in the present time and in the future. Several innovative visionaries, thinkers, organizational leaders, consultants, and scholars in transportation and high-speed rail related area have presented their recommendations in this book, which would facilitate change a vision into reality. High Speed Rail is certainly the "LINK" to the future for the United States and the world!

Raj Selladurai
Indiana University Northwest, USA

Peggy Daniels Lee
Kelley School of Business, Indiana University, USA

George VandeWerken
Providence Bank, USA

REFERENCE

Indiana, High Speed Rail: 'LINK' to the Future. (2013). *The South Shore Journal*, *5*, 171-182.

Acknowledgment

I would like to thank my co-editors Dr. Peggy Daniels Lee and Mr. George VandeWerken for all their advice, help, recommendations, and support with this book. It was indeed a wonderful academic experience working with them on this book project. I wish to acknowledge my dean, Dr. Anna Rominger, and Dr. Ken Carow, dean at the Kelley School of Business, Indianapolis for their encouragement and support of my research interests and efforts in high-speed rail; and to Dean Carow for kindly referring me to one of his faculty members, Peggy Lee, to make up our book editorial team.

I would also like to thank my friend Dennis Hodges, Vice President, Business Relations with Indiana Passenger Rail Alliance. Dennis, a pioneer in the high-speed rail movement especially in Indiana, has been a major inspirational source for my passion and interest in high-speed rail research and has helped immensely through his networking contacts on various aspects of this book project.

I would like to thank my wife, Revathi Rebecca, my daughter Roshini Isabell, and my son Roshan Isaac for their total, continually growing love, patience, support, and encouragement.

Above all, I would like to thank my Lord Jesus for everything, and I wish to dedicate this book to Him all for His glory....Colossians 3:17.

Raj Selladurai
Indiana University Northwest, USA

The chapters that I have helped to write for this book would not have happened without the encouragement and support of a number of colleagues and friends. I am grateful to David Harvey, a banking colleague and friend, who encouraged my initial interest in public transportation. In addition, Dennis Hodges, the founding member of the Indiana High Speed Rail Association (now known as the Indiana Passenger Rail Alliance) reached out and invited me to join that organization and become a board member. That experience provided greater exposure to the fascinating topic of public transportation and whetted my appetite for more. Finally, I am especially grateful for the encouragement of Raj Selladurai, my co-author and friend, whose enthusiasm for this topic and desire to capture new ways of thinking about familiar transportation topics has been a catalytic inspiration.

George VandeWerken
Providence Bank, USA

Many thanks to my co-authors for giving me the opportunity to be a part of a great editorial team. Any work would not be possible without the assistance of those who support us in our endeavors. Thus, I am also grateful to Donald McCray, who as a freshman at the Kelley School of Business Indianapolis, who worked on the literature review; the Kelley Indianapolis faculty secretaries; Kenneth Carow, Associate Dean for Research and Programs at Kelley Indianapolis, for getting me involved in the project initially; and to all of my colleagues at Kelley Indianapolis for their constant support and encouragement.

Peggy Daniels Lee
Kelley School of Business, Indiana University, USA

Section 1

Chapter 1
California High-Speed Rail:
A Transformative Investment in California's Future

Jeff Morales
California High Speed Rail Authority, USA

ABSTRACT

California has started construction on the first true high-speed rail system in the United States. In this chapter, the CEO of the California High-Speed Rail Authority describes how the state's historic investment in new high-speed rail infrastructure will benefit Californians. Benefits include job creation, statewide modernization of existing local rail lines, short and long-term reductions in greenhouse gas emissions, preservation of agricultural land and environmentally sensitive habitats, advanced clean and green construction practices and technologies, a commitment to 100% renewable energy use, and assistance to partner cities on transit-oriented planning for high-speed rail station areas.

INTRODUCTION

With the official groundbreaking ceremony having been held on January 6, 2015 and construction currently underway, California continues to lead the country in its effort to bring high-speed rail to America. California's high-speed rail system, which will initially connect San Francisco to Los Angeles, before adding on sections

DOI: 10.4018/978-1-5225-0102-2.ch001

to San Diego and Sacramento, will usher in an era of clean transportation and economic growth for the state's economic hubs and fastest growing regions. California as a state is perfect for a modern high-speed rail system. The state's population is projected to grow beyond 50 million people by 2050 – an increase of over 30% from the state's current 38 million (California Department of Finance, n. d.).

Demographically, California has seen a shift in how people view public transportation, with Millennials showing a preference for alternative sources of transportation such as high-speed rail instead of traditional interstate travel options like cars and planes. Geographically, California's high-speed rail system is bookended by two major metropolitan population centers (San Francisco Bay Area and Los Angeles/ Anaheim) and an alignment that connects millions of Central valley residents to these centers. This connectivity via high-speed rail will serve to further the environmental and economic development goals of the state while connecting people, services, and goods. All these benefits are made possible through the most ambitious and needed infrastructure program in the country.

California has seven of the nation's ten most polluted cities in terms of air quality (American Lung Association 2013). The state also has six of the 30 most congested traffic areas in the United States as well as the country's busiest short haul air market – Los Angeles to San Francisco Bay Area. One in five (20 percent) flights between Los Angeles to San Francisco is delayed by about an hour, which is about the same amount of time it takes to fly from airport to airport. High-speed rail in California can help those problems. If California were to follow the same infrastructure funding formula that America has used for the past 50 years, the price tag would be twice as much as high-speed rail, and require over 4,300 new highway lane miles, 115 additional airport gates, and four new runway terminal. The cost to California if it were to stick with the status quo of more roads and airports/runways would be an estimated $158 billion in construction alone, with another $133 billion in operations and maintenance over the next 50-years. With high-speed rail in place, California can reduce congestion on highways and in airports; help improve air quality to meet the state's landmark greenhouse gas reduction goals; push the envelope in terms of the most cutting edge, environmentally sound construction practices and land-use mitigation solutions; and put thousands of people back to work, especially in the Central Valley, where the recession caused unemployment numbers to double the state's average.

BACKGROUND

California's pursuit of a high-speed rail system started three and a half decades ago, in 1981. At that point, California looked into a Southern California high-speed rail corridor which would have been built through a partnership with Japanese high-speed rail firms. Over a decade later, in the mid-1990s, the state's growing population put increasing strain on existing transportation systems, causing officials to look at implementing high-speed rail on a statewide level. In 1994, the Federal High-Speed Rail Development Act, authored by then Representative (now current High-Speed Rail Authority Board Member) Lynn Schenk, identified California as one of five corridors in the nation for high-speed rail planning. During that same timeframe, the California Legislature (Legislature) created the Intercity High-Speed Rail Commission and charged it with exploring the feasibility of a California high-speed rail system. In 1996, the Commission issued a report that confirmed the feasibility of California's high-speed rail project.

Also in 1996, the Legislature created the California High-Speed Rail Authority (Authority) to plan, prepare, and design the construction of a system to connect the state's major metropolitan areas. In 2000, the Authority released its first business plan; two years later, Senate Bill (SB) 1856 (Costa) was passed by the Legislature, authorizing a $9.95 billion bond measure to finance the system – submission of this measure to the voters was delayed until 2008. As part of its planning responsibilities, the Authority – together with the Federal Railroad Administration (FRA) – issued a Draft Program-Wide Environmental Impact Report /Environmental Impact Statement (EIR/EIS) as required by the California Environmental Quality Act (CEQA). The EIR/EIS described the system and its potential impacts on a statewide scale. Through that exercise and process, the Authority received upwards of 2,000 comments submitted by government agencies and members of the public. These comments were used by the Authority to determine the preferred corridors and station locations for the future system.

The bond measure authored by then-State Senator Costa was put to voters in 2008, a moment that many in the state believe to be the beginning point for high-speed rail. The bond measure was approved by voters, making California the first state in the nation with a voter-approved financing mechanism to develop high-speed rail. In 2009, $8 billion in federal funds was made available nationwide as part of the American Recovery and Reinvestment Act (ARRA), a measure passed to stimulate the national economy, create jobs, and facilitate new rail manufacturing enterprises. The ARRA funds demonstrated the national commitment to developing high-speed rail, as described by the U.S. Department of Transportation's plan, "A Vision of High-Speed Rail in America."

California pursued the ARRA funds and other available federal funds, successfully securing $3.3 billion for planning, environmental work, and construction.

Phase 1 of the high-speed rail system, as established by the Safe, Reliable High-Speed Passenger Train Bond Act for the 21st Century (Bond Act), is the North-South corridor between San Francisco's Transbay Terminal and Los Angeles' Union Station/Anaheim. The Authority's Board of Directors (Board) evaluated segment proposals in November of 2011, with unsubsidized service being among many other criteria for Initial Operating Segments (IOS). Two IOS choices were selected from this evaluation, both of which include multiple stations as required by statute:

- IOS-North is the portion of Phase 1 corridor from San Jose to Bakersfield; and
- IOS-South is the portion of Phase 1 corridor between Merced and the San Fernando Valley.

The first 130-mile construction section is located on both IOS-North and IOS-South.

When the Board adopted the 2012 Business Plan, it selected the IOS-South to advanced first, closing the rail gap between Northern and Southern California (there are currently no passenger tracks that cross the Tehachapi mountain range) and providing an early connection into Los Angeles through a high-speed rail/Metrolink (existing rail service) connection at Palmdale.

During Governor Edmund G. Brown Jr.'s 2012 State of the State address, he highlighted the benefits of high-speed rail and declared the project a priority for his administration, continuing the support of his predecessor, Governor Arnold Schwarzenegger.

Another important milestone of 2012 was the development and adoption of the 2012 Revised Business Plan. The revised plan broke away from preliminary ideas and laid a solid framework for implementing the California high-speed rail system as part of a broader statewide rail modernization program, in concert with other state, regional, and local rail investments. The Legislature passed Senate Bill 1029 (Budget Act of 2012) that same year allocating nearly $8 billion in federal and state funds towards major investments in California's infrastructure: construction of a true high-speed rail system running through California's Central Valley and 15 other bookend and connectivity projects to upgrade existing rail infrastructure throughout the state in preparation for high-speed rail.

In the 2014 state budget negotiations, the Legislature and the Governor's Office settled on an ongoing funding source for high-speed. Under the budget agreement, the Authority received a one-time funding package of $250 million for Fiscal year 2014-2015 from California Air Resources Board's Cap-and-Trade auction proceeds. Under the same budget agreement, starting in Fiscal Year 2015-2016, the Authority will receive 25 percent of the total Cap-and-Trade auction proceeds in perpetuity.

CAP-AND-TRADE CHANGES EVERYTHING

The unprecedented allocation of an ongoing funding source with Cap-and-Trade proceeds was a game changer for California's high-speed rail program. For the first time, the Legislators allocated a steady stream of funding for transit and rail. This allocation showed outside investors that the state is serious about high-speed rail and is committed to investing future funds to make sure it happens. Prior to that vote, letters of support from industry leaders such as AECOM, Siemens, Railgrup, Acciona and Astaldi expressed a commitment to high-speed rail and future private sector involvement.

The Cap-and-Trade proceeds allow the Authority to look at accelerating the construction of California's high-speed rail program through concurrent construction projects. In addition to the first construction segment in the Central Valley, the Authority is looking at ways to move forward with construction in the Palmdale to Burbank and Burbank to Los Angeles corridors of the program. The Authority has started the process of reviewing and analyzing alignment options between these communities by hosting public meetings and asking for feedback to several proposed alignments.

In Southern California, the city of Palmdale has been designated as the site for the "golden spike." Palmdale, a community of over 150,000 people, is located 64 miles outside of Los Angeles, but the commute to Southern California's economic center can be as long as three hours in bumper-to-bumper traffic and two hours via existing rail. Bringing high-speed rail to Palmdale connects California's Central Valley to Southern California for the first time via passenger rail. Accelerating the Los Angeles to Palmdale connection will allow residents to get to and from L.A. in as little as 30 minutes. In addition to being a home for California's high-speed rail system, XpressWest, a privately-funded program looking to extend high-speed rail service from Las Vegas to Southern California, has said they want to make Palmdale their terminus for high-speed rail service. If successful, this service would allow passengers to use California's high-speed rail service from anywhere in the system to connect to Las Vegas.

WHY HIGH-SPEED RAIL WILL WORK IN CALIFORNIA

It Complements and Improves Existing Transportation Infrastructure

Significant demand for intercity travel in California has maximized the capacity of the state's existing transportation systems. Auto congestion within the state drains approximately \$19.5 billion in wasted time and fuel from the economy every year. California's intercity rail lines are seeing increased ridership: the Pacific Surfliner (San Luis Obispo to San Diego), Capitol Corridor (Sacramento to San Jose) and San Joaquin (Central Valley) corridors (Amtrak), rank second, third and fifth in the nation respectively, and ridership has ballooned on these three lines by 256 percent (Capitol Corridor), 66 percent (San Joaquin), and 61 percent (Pacific Surfliner), between 1997 and 2012.

The state's major airports are also hitting limitations of capacity. Each year, more than eight million passengers fly between the Los Angeles and Bay Area airports. Los Angeles International Airport (LAX), which hosted about 62 million passengers in 2012, has few, if any expansion options. Los Angeles World Airports (LAWA), the City of Los Angeles department responsible for three airports including LAX, expects to reach its capacity of 78.9 million passengers per year in 2022, years ahead of the completion schedule for Phase 1 of the high-speed rail system.

Meanwhile in the Bay Area, the largest airport serving the Bay Area and northern California, San Francisco International Airport (SFO), hosted just over 42 million passengers in 2010. By 2035 that number is projected to grow to over 100 million – a 138% increase. Located by the environmentally sensitive San Francisco Bay makes SFO an unlikely candidate for expansion or additional runway capacity to meet the needs of these new passengers. In evaluating various options for demand management, SFO concluded that high-speed rail was the most effective measure for complimenting its services and optimizing existing capacity.

High-speed rail's implementation would allow travelers throughout the state a more environmentally friendly alternative to planes and cars, while also providing an on-time service with a comparable length of travel time.

Modernization at the Bookends

Construction work on California's high-speed rail system started in the Central Valley in 2012. This allowed construction companies to prepare for building the backbone of the first true high-speed rail system in the United States. Connectivity and bookend enhancements have progressed at the same time, including clearance of environmental planning of the proposed electrification of the Caltrain corridor (the segment of the system where the high-speed rail will share tracks with the South Bay's Caltrain system), construction of the Central Subway project in San Francisco, and the identification and advancement of a number of projects in Southern California, including improvements to San Diego's Blue Line Transit System, and the Regional Connector in Los Angeles.

In the near term, these projects will strengthen and improve existing rail networks, which will yield early mobility benefits throughout California. As the high-speed rail system is eventually phased in throughout the state, these completed projects will enhance the high-speed rail system's utility by providing seamless connections with local and regional rail systems.

Fast Connections between Economic Centers will Enhance California's Economic Competitiveness

California is an economic powerhouse. California is the seventh largest economy in the world. It is home to a range of small and large businesses and economic clusters, world-renowned universities and cutting-edge start-ups. Connecting these economic clusters with fast, direct downtown-to-downtown service will help spur even greater collaboration and innovation.

Currently, California's Central Valley is separated from the state's two largest economic hubs because the state lacks viable transportation options. High-speed rail will connect the Central Valley's growing population centers to the state's economic and business centers for the first time. Connectivity through high-speed rail will allow companies in Los Angeles, Silicon Valley, and San Francisco to invest in the growth of cities like Merced, Fresno, and Bakersfield – where land for industry is available and cost of living for employees is cheaper in comparison. California's high-speed rail program is looking at data from existing high-speed rail systems around the world to help local governments prepare for the opportunities that connectivity brings.

High-Speed Rail Brings Economic and Environmental Benefits to California's Central Valley

While the Central Valley is blessed with some of the richest agricultural land in the world, unemployment remains stuck in the double digits. Air quality is a major concern, with every major city in the San Joaquin Valley ranking among the top-10 for worst air in the nation. The American Lung Association has estimated that one in four people in the Central Valley will have some sort of respiratory illness because of the air quality. Connecting the Central Valley to the broader state economy with a clean transportation system will support both economic diversification and efforts to reduce pollution.

Building and operating the high-speed rail system will directly employ thousands of Californians, while indirectly generating tens of thousands more jobs throughout the larger economy. Construction on the first segment between Madera and Bakersfield is projected to create thousands of jobs over the next five years. Also, permanent jobs – train operators, maintenance yard workers, stations managers and others – will be created to operate and maintain the system. For example, according to the bottom-up estimate of the labor required in the O&M estimates to run the system, the Initial Operating System between Merced and the San Fernando Valley is expected to directly employ almost 1,500 people full time in operations.

High-Speed Rail Helps California Meet Environmental Goals through Sustainable Practices

California's high-speed rail program is committed to using renewable energy for powering the system, net zero greenhouse gas (GHG) emissions in construction, and recycling 100 percent of the steel scrap and concrete refuse generated in project construction. This commitment already has been put into practice with the Construction Package 1 (CP 1) design-build contract. Net-zero energy operations can be achieved by procuring or producing enough renewable energy to offset the amount of energy the system takes from the state's power grid to operate trains and facilities. The combined result of green practices and the resulting shift from autos and planes to the high-speed rail system means that the system is projected to save 1.7 million barrels of oil annually starting in 2030, which is the equivalent of all of the electricity used in over 1 million homes in a month.

Prior to 2020, Proposition 1A investments in urban transit systems and rail modernization projects like the Caltrain electrification project will result in tens of thousands of tons of reductions in GHG emissions. California leads the nation in establishing policies to reduce GHG emissions. In 2005, former Governor Schwarzenegger signed an Executive Order directing that GHG emissions be reduced by 80 percent of 1990 levels by 2050 and in 2006; the Legislature passed AB 32, the Global Warming Solutions Act, which further directs the state to reduce statewide emissions to 1990 levels by 2020. Since ARB's 2008 Scoping Plan, the high-speed rail system has been listed as a measure to achieve the state's GHG emission reduction goals.

Using California ARB and U.S. Environmental Protection Agency data sets, models, and methodologies, the Authority estimates that:

- In 2022, when the IOS is up and running, the resulting GHG reductions will be over 130,000 metric tons of carbon dioxide in the first year, which is the equivalent of over 23,000 personal vehicles taken off the road.
- Between 2022 and 2040, the cumulative reduction of carbon dioxide is estimated to be over 10 million metric tons.
- Prior to 2020, investments in early rail modernization projects will yield reduced GHG emissions, including, for example, Caltrain electrification, which will result in at least 68,000 tons of reductions.

In addition, the CARB and other state planning agencies view high-speed rail as an investment that can be a catalyst to effective land use planning and transit oriented development, which will provide enormous additional savings of GHGs.

Millennials are Trending towards Alternative Transportation Options

The U.S. population is in the middle of a demographic shift. The Millennials, or Generation "Y," born between 1983 and 2000, comprise the largest generation in the nation with over 79 million people. Millennials have grown up during an era of rapid technological advances and urban growth and their preferences are increasingly driving larger cultural, consumption, and travel trends.

Millennials are opting for urban, walkable neighborhoods and apartments or condos. Instead of relying on cars for all or most of their trips, Millennials are taking public transit, riding Amtrak, biking, walking, and using other services. A recent study on Millennials and mobility, conducted by the American Public Transportation Association (APTA), found that they are attracted to communities that offer a variety of transportation choices.

Millennials are the first generation to have grown up with smart phones and mobile devices that make it easier to plan trips, obtain real-time travel information, and even pay fares (Dutzik and Baxandall, 2013). Technology allows Millennials to use the time they spend waiting for or traveling on planes or trains more productively.

All of these factors underscore the fact that Millennials are driving less than previous generations and are choosing to use public transit and other non-auto modes for some or all of their trips. Per capita vehicle-miles traveled by 16-34 years olds has decreased by almost 25 percent between 2001 and 2009 (from 10,300 miles to 7,900 miles per capita). The vehicle miles traveled per person in 2011 in California was 8,511 miles, a 6.7 percent decrease from 2005, although this may be attributed to the recession (Baxandall, 2013).

This reflects and is consistent with large increases in rail ridership, nationwide and in California. Between 2011 and 2012, ridership increased by 18.5 percent on LA Metro's light rail system, by 7.8 percent on BART's heavy rail system and by 13 percent on Caltrain in the Bay Area.

These trends mean that California High-Speed Rail will be coming online when demand for rail will be at its highest in decades and continuing to grow. In a 2012 APTA poll on high-speed trains, 74 percent of respondents between 18-24 years old stated they were likely to use high-speed rail service for business or leisure travel (American Public Transportation Association, 2012). As high-speed rail becomes part of California's transportation network over the next several decades, the largest and most public transit-oriented generation will be getting into its prime traveling years.

WHERE WE ARE, WHERE WE'RE GOING

San Francisco to San Jose:

Construction on California's high-speed rail system is occurring on multiple fronts. In Downtown San Francisco, the station which will be the northern terminus for Phase 1, San Francisco's Transbay Transit Center (TTC) is well underway. As a first step towards high-speed rail service in the San Francisco Bay Area, the TTC has completed construction on the area of the station which will house the train box – the underground portion of the TTC that will be dedicated to Caltrain and high-speed rail. Development of the Terminal building has begun along with construction of what will be the tallest structure in San Francisco, the Salesforce Tower. The tower will soar to 1,070 feet, making it the seventh-tallest building in the country as well as an anchor of development for the area around the northern terminus. Back at the ground level of the tower, the TTC will service eight San Francisco Bay Area counties through 11 different transit systems, including: AC Transit, Amtrak, BART, Caltrain, Golden Gate Transit, Greyhound, SF MUNU, SamTrans, WestCAT Lynx, and Paratransit.

California's high-speed rail program is benefitting Bay Area commuters in the immediate future by partnering with Caltrain on their electrification goals. Electrification of the Caltrain corridor, from San Jose to San Francisco, is made possible through high-speed rail funding. As part of SB 1029, Caltrain was provided $705 million to install and enable the replacement of diesel trains and connect the system with high-speed rail. This will allow Caltrain to provide electrified service as soon as 2021 and reduce their GHG emissions by approximately 68,000 tons of carbon dioxide (CO_2) per year.

SB 1029 also appropriated $42 million, which is combined with contributions from BART and the Santa Clara Valley Transportation Authority, for the design, installation, testing, training and warranty for an intelligent network of signals, sensors, train-tracking technology and computer systems on the Caltrain Corridor as part of Caltrain's advanced signaling system. Construction of this system, more commonly referred to as Positive Train Control (PTC), is underway as required by federal regulation and will allow trains to travel safely at higher speeds and with greater frequency. The Authority forecasts completion of this work within the next couple of years.

San Jose to Merced

In 2014, the Authority and the City of Gilroy entered into a station-area funding agreement. These agreements allow cities along the high-speed rail alignment to initiate the planning process for a potential high-speed rail station. The City of Gilroy is looking into the effects of a high-speed rail station located in Downtown Gilroy. The station-area funding agreement will allow the Authority and city to engage in public outreach meetings designed to seek input on various options for the planned stations. The Authority is expected to enter into a similar station-area funding agreement with the City of San Jose.

Merced to Bakersfield

Work on CP 1, the first 29-mile stretch of California's high-speed rail system, has increased significantly since summer of 2014. In June 2014, Tutor-Perini/Zachry/Parsons (TPZP), the Authority's design-build contractor, conducted a series of statnamic load tests. These mini-explosions tested the loadbearing columns that will be used for viaducts serving as high-speed rail bridges over the Fresno River. Shortly after the testing, the Authority began demolition of existing structures along the high-speed rail alignment in preparation of large scale construction.

As construction continues, the Authority and its local partners have worked together to direct potential job seekers to training opportunities for high-speed rail construction work. This includes a series of six-week training sessions to qualify workers for apprenticeship programs to become engineers and laborers.

Work on the Fresno to Bakersfield portion of the system moved forward in May 2014 when the Board certified the Final EIR/EIS for the Fresno to Bakersfield project section. In June 2014, the FRA issued their Record of Decision (ROD) under the National Environmental Policy Act (NEPA), followed by the Surface Transportation Board (STB) issuing a concurrence on the document and authorization to proceed with construction in August 2014. This series of approvals allowed the Authority to begin the procurement process for Construction Package 2-3 (CP 2-3), a 60-mile construction section that begins in Fresno and ends just north of the Tulare-Kern County lines. In June 2015, the Authority executed a contract with international construction firm Dragados, along with American firms Flatiron Construction Corp. for the design-build of CP 2-3. The consortium of Dragados/Flatiron came in with a bid well below the Authority's estimates of $1.5-$2 billion while also scoring the highest on the technical proposal portion of the bid.

Palmdale to Los Angeles

With the ongoing commitment of Cap-and-Trade proceeds to the high-speed rail program, the Authority has designated Palmdale to be the "golden spike" of California's high-speed rail system. The Authority is poised to advance the planning and construction of multiple segments concurrently, resulting in a number of key environmental and mobility benefits. In addition to the work being done in California's Central Valley, the Authority has specifically identified the newly-formed Palmdale to Burbank project section as a key segment that could potentially be accelerated by Cap and Trade proceeds as they become available.

The Palmdale to Burbank project section will travel from the Palmdale Transportation Center southward to the Burbank Airport Station, approximately 45-miles. The Burbank to Los Angeles project section would travel from the Burbank Airport Station following the existing rail corridor to Los Angeles Union Station in Downtown Los Angeles, a 15-mile route. As a result of local stakeholder input, the Authority is studying a more direct route from Palmdale to Burbank, one that would tunnel through the San Gabriel Mountains. This route may be shorter and avoid impacts to communities along the original corridor.

In addition to the work the to connect the Antelope and San Fernando Valleys, the Authority continues to advance near-term investments and partnerships to deliver regional mobility projects like the Southern California Regional Interconnector Project (SCRIP) that will materially improve the Los Angeles-San Diego-San Luis Obispo (LOSSAN) corridor service for Metrolink and Amtrak, as well as the movement of goods and services.

Los Angeles to Anaheim

The City of Anaheim recently opened California's first multi-modal station built with high-speed rail in mind. The Anaheim Regional Transportation Intermodal Center (ARTIC) – a multi-modal and intermodal hub for Orange County – opened in December of 2014 and includes services such as Amtrak, Metrolink, OCTA bus, and Anaheim Resort Transportation as well as shuttles, taxis, and charter busses.

The Authority continues to refine alignment alternatives appropriate for the urban rail corridor from Los Angeles to Anaheim. This approach incorporates the principles of an integrated passenger rail network described in the Authority's 2014 Business Plan. As the Authority works with local stakeholders on high-speed rail implementation, partnerships with local transit agencies advance near-term investments designed to deliver regional mobility projects like the grade separations and Amtrak and Metrolink service improvements.

CONCLUSION

The transformative things that make California great – the Golden Gate Bridge, the world renowned University of California system, the California Water system – were only achieved because a select few saw that the needs of the future were worth the political fights of the present. Those who wanted to build the Golden Gate Bridge had to face over 2,000 lawsuits brought before the courts to stop the construction, but the builders prevailed, work continued, and California has the world's most iconic bridge because of it. What would the San Francisco Bay Area be without it?

The University of California is the world's most renowned set of public institutions, yet the creation of the system – as well as the California water delivery system – only passed through the legislature by single vote margins. California would not be the world's eighth largest economy without these two programs in place, but the political victories to create them barely made it through the legislature.

California's high-speed rail program is following in these footsteps. While there is opposition to the program, the country's first true high-speed system continues to move forward. Creating high-speed rail in California will provide economic benefits and connect the state in the same way that the University of California and state water systems have. In addition to the economic benefits, high-speed rail will give Californians an environmentally friendly option for travel, and allow passengers convenient one-seat travel from Downtown Los Angeles to Downtown San Francisco (with San Diego and Sacramento added later) in the same time it takes to fly using congested airports for these short regional trips. The train will provide California will usher in California's next golden rush. Connectivity, environmentally friendly transportation, smart and sustainable city design, and new iconic high-speed rail stations will all be developed because of the high-speed rail program. Fifty years from now, California's high-speed rail program will be mentioned in the same sentences as the Golden Gate Bridge, University of California system, and state water project when talking about the transformative things that make California the leader of America's economic and environmental policies.

REFERENCES

American Lung Association (ALA). (2013). Most Polluted Cities. Retrieved from http://www.stateoftheair.org/2013/city-rankings/most-polluted-cities.html

American Public Transportation Association. (2012, May). APTA High-Speed Train Survey.

Baxandall, P. (2013, August). *Moving off the Road: A State-by-State Analysis of the National Decline in Driving*. U.S. PIRG Education Fund.

California Department of Finance. (n. d.). Report P-1: Summary Population Projections by Race/Ethnicity and by Major Age Groups. Retrieved from http://www.dof.ca.gov/research/demographic/reports/projections/P-1/

Dutzik, T., & Baxandall, P. (2013). A New Direction Our Changing Relationship with Driving and the Implications for America's Future. U.S. PIRG Education Fund, Frontier Group.

KEY TERMS AND DEFINITIONS

High-Speed Rail: A method of rail transportation that uses electricity to operate significantly faster than traditional rail counterparts.

Millennials: A person born between 1982 and 1994.

Positive Train Control: Using satellite technology and a series of checkpoints to monitor and control high-speed trains as a way to enhance safety and efficiency.

Trainsets: Single level Electric Multiple Units capable of operating in revenue service at speeds of 220 mph.

Transit-Oriented Development: Residential and commercial construction and planning that intends to tie into existing and proposed transit as well as encourage walking, running, biking.

Chapter 2
California's "Fast-Track" to High-Speed Rail:
The Early Challenges and Ultimate Success of the California High-Speed Rail Project

Rod Diridon Sr.
Mineta Transportation Institute, USA

Ben Tripousis
California High-Speed Rail Authority, USA

ABSTRACT

For over 50 years, after both public and private studies confirmed California's need for high-speed rail, nothing happened. The rest of the US had a similarly disappointing experience when attempting to break into the rapidly growing family of nations with operating high-speed trains. Yet, over the past year, California has contracted over $2.5 billion in high-speed rail civil work, is preparing RFPs for three times that amount, and has secured commitments for nearly $25 billion in additional federal and state funding over the next 30 years. Private investors are lining up for the nation's first high-speed rail public-private partnership to complete this $68 billion, 540 mile project. This chapter describes the unprecedented technical excellence, political courage, and visionary voter support that created that high-speed rail breakthrough for California and the nation.

DOI: 10.4018/978-1-5225-0102-2.ch002

INTRODUCTION

The US has been denied the advantages of high-speed rail now enjoyed by all of the other industrialized nations of the world. Japan's Shinkansen has been operating for over 50 years, while the French TGV began over 30 years ago. After carrying billions of riders, neither system has experienced even one fatality from operations. China, in just 15 years of construction, has completed over 8,000 miles of a 230-mile-per-hour system, with more under construction. Spain, an economically challenged country, has built over 2,000 miles of high-speed rail, creating good jobs and promoting economic recovery. The success stories are legion, with few if any failures. The tried-and-true adage is that the country that moves people to work and product to market most efficiently wins the geoeconomic competition. On that basis, the US, in its reliance on gridlocked highways and inefficient short-hop air travel in the congested major metropolitan regions, is losing that competition and has little hope of catching up. And, unlike Japan and France, which are in compliance with the historic Kyoto accords, the US has no prospect of meeting those now-obsolete but still internationally desired guidelines for combating climate change. The US is relying too heavily on carbon-intensive power and shunning high-speed rail technology, the linchpin of the other electrically powered international transportation systems.

This chapter carries the reader step-by-step through California's many faltering steps to the current successful high-speed rail construction project. The confounding experiences are shared along with the remarkable examples of insight, courage, and public support. The homily shared by both Californians and others, some with pride and others with fear, is that as California goes so goes the nation. Indeed, California will now be a high-speed rail state. Many in California and across the country hope that the homily is true.

Background

The explosion of rail service in the late 1800s opened up and molded California. Prior to the introduction of the Transcontinental Railroad, regarded by many as the greatest project ever attempted, the trickle of hardy easterners to California barely changed the Spanish Colonial culture of the region that had recently been ceded to the United States by Mexico. But the sleepy social and economic system created by the Catholic Franciscans and the Spanish grandees was about to be awakened with the clang of hammers on steel and the urgent, strident, impatient sound of steam!

Tens of thousands of Chinese, Irish, and Italian laborers earned their right to US citizenship while binding the country with a new "high-speed" rail system that could travel as fast as 25 miles per hour, then 50, 75, and – on a good day with no inspectors watching – over 100 miles per hour using high-pressure steam produced by wood, then coal, then #3 bunker fuel. In 1950 another wave of innovation caught the old steam "hogheads" by surprise. Electricity began to be created by massive diesel-fueled internal combustion engines. A new system, arguably more efficient and flexible but no faster, replaced the water tanks, roundhouses, and massive repair facilities, putting millions of steam rail specialists, who had supported the grand old locomotives for nearly 100 years, out of work.

But the speed seemed to have topped out at the 79 miles-per-hour maximum that the Federal Railroad Administration deemed safe for most main line railroads. Then Japan, in a post-WWII embrace of capitalism, decided to create the Shinkansen – "happy wind" – bullet train. Using high-voltage electricity conducted from a charged wire via the train's overhead catenary, the 135-mile-per-hour Japanese high-speed rail system introduced in October 1964 captivated the world. Note, the modern Shinkansen and most other major high speed train are now capable of sustain speeds in excess of 200 miles-per-hour. It immediately became both an economic success and declaration that Japan had returned as one of the world's top manufacturing powerhouses. France soon followed with its own high-speed train, à Grande Vitesse (TGV), in 1981, followed in quick succession by Germany, Spain, Italy, Korea, Taiwan, China – the current world leader in high-speed rail track miles, and many others. Indeed, every industrialized and most emerging nations – even a few developing countries – now have high-speed rail systems (300 KPH or 187 MPH is the international HSR standard) in operation or under construction. All, that is, except for the United States.

While the US government began scrambling to catch up, railroads were under pressure to move people and product faster and more efficiently due to competition from the rapidly evolving airline industry. In the late 1950s, the Association of American Railroads (AAR), spurred by Japan's successful tests of the Shinkansen, called for the nation's class-one railroads to examine the feasibility of high-speed routes serving major US markets. California's venerable Southern Pacific Railroad (SPRR), since acquired by Union Pacific (UPRR), assigned the task to James C. Strong, a bright young railroad design engineer recently graduated from the University of California at Berkeley.

Quoting from a 2011 letter in the Mineta Transportation Institute's (MTI) archives by Mr. Strong:

A few salient details about the route are: 1) All plotting was done using USGS maps. 2) The SPRR route was used where possible and began in San Francisco south to Tunnel 14 in South San Francisco. 3) From Tunnel 14 the route diverged east across highway 101 to the bay. 4) The route continued in San Francisco Bay on a long trestle on pilings 5) The route stayed off shore in the shallow bay until reaching Milpitas in north Santa Clara County. 6) Proceeding south via San Jose and Gilroy on the SPRR tracks the route diverged along SP's Hollister Line south east along the Salinas River to Tres Pinos. 7) Following one side of the San Andreas Fault, the route crossed Pinoche Pass. 8) The San Joaquin Valley section was flat and straight. 9) The route passed near Coalinga, Taft, and Wheeler Ridge. 10) A 28-mile tunnel was then required leading to Castaic (striking oil was feared in that section). 11) An added five-mile tunnel was needed to traverse from Santa Paula Valley to San Fernando Valley. 12) The last section into the LAUP Station used the LAD/W&P and SCE high voltage transmission rights of way and was adjacent to the Los Angeles River at the south end. (Mineta Transportation Institute, 2016)

Mr. Strong's route is slightly shorter than the route chosen by the California High-Speed Rail Authority, which avoids the long bay trestle, serves the rapidly urbanizing north Central Valley/Fresno area, and minimizes the tunnels under the Tehachapi Mountains into the Los Angeles Basin.

Unfortunately, since the federal government had no construction funding at that time, AAR stopped promoting the idea, and SPRR felt that, though high-speed rail was eminently possible and economically viable, the freight railroads had no need for that kind of speed. The bulk of the passenger service was already lost to such emerging competitors as Pacific Southwest Airlines, connecting the state's largest metro areas, and the highway systems were not yet terribly congested. So Mr. Strong's study was shelved (though is still available in the MTI and AAR archives) and high-speed rail languished in California and the nation for decades (Mineta Transportation Institute, 2015).

PROGRESS ABROAD MAKES THE FUTURE CLEAR

North America, bound by culture and politics to domestically obtained and then-inexpensive petroleum power, continued to lumber along at no more than 79 mph, burning diesel fuel and creating a level of noise, vibrations, air pollution, and general congestion that was becoming an embarrassment when compared to the rest of the world.

That comparison became more prevalent as growing affluence, especially in California, allowed extensive international travel. Governor Edmund G. "Pat" Brown Sr. formalized the comparative opportunity by creating the California Institute for Technology Exchange (CITE) in the 1960s encouraging the exposure of state governmental leaders to new technology throughout the world. Such later luminaries as Jim Costa, Rod Diridon, Richard Katz, Quentin Kopp, Rebecca Morgan, Mehdi Morshed, Lynn Schenk, and many others participated in CITE-sponsored high-speed rail inspection trips to Japan and Europe in the 1970s and 1980s, sowing the seeds of interest that resulted in Assembly Bill 971 in 1987. That measure created the High Speed Rail Committee that organized the California High Speed Rail Commission, which later evolved into the 1996 legislation creating the current California High-Speed Rail Authority (CHSRA).

In 1974, then-Governor Edmund G. "Jerry" Brown, Jr., Pat Brown's son, signed related legislation calling for a feasibility study of developing high-speed rail in California. However, it was not until 1993 that then-Congresswoman Lynn Schenk, often referred to as the "mother of California's high-speed rail," co-authored H.R. 4867, the High-Speed Rail Development Act of 1994, the first federal law authorizing the planning and development of high-speed rail.

The current California High-Speed Rail Authority (CHSRA) was established in 1996 by State Senate Bill 1420 (Chapter 796 of California Statutes). The CHSRA is legally tasked to design, build, operate and maintain a high-speed train for the state. Paradoxically, then-Senator Kopp authored the bill but in recent years has opposed plans to create the system. The legislation was signed into law by then-Governor Pete Wilson, who supported high-speed rail with some concerns. His initial five appointees to the CHSRA board reflected his reservations and were, at best, lukewarm to the idea of high-speed rail in California. Chair Michael Tennenbaum, with Executive Director Mehdi Morshed (a 20-year leader of the State Senate Transportation Committee's staff and a strong HSR supporter) established the obligatory relationship with the Federal Railroad Administration for oversight. In 1997, the CHSRA advertised for bid the Authority's formal "concept study." Noted infrastructure planning and development firm Parsons Brinckerhoff won that bid and has continued as the project manager since that time (California High Speed Rail Authority, 2015).

INCONSISTENT FUNDING, DELAYS, AND CHALLENGES

For the next decade, funding for the Authority was provided annually from the State's general fund. An occasional federal study grant was received, most often with the assistance of US Senators Barbara Boxer and Dianne Feinstein, both strong supporters of high-speed rail. Unfortunately, the annual funding that was appropriated was in the tens of millions of dollars, not the hundreds of millions needed to move the vast project forward. Having to rely on annual budgets precluded long-term planning and resulted in frequent work stoppages as budget approvals came late, requests were cut, and staffing was restricted.

In 2000, a concept study draft was submitted to the CHSRA board for public hearing and was approved unanimously. The concept clearly identified the technical and financial viability of the roughly 790-mile system connecting Sacramento, via the Central Valley, to Los Angeles and San Diego and with the Bay Area via a wye just north of Fresno at Chowchilla then through the Pacheco Pass via Gilroy to San Jose's job-rich Silicon Valley and finally to San Francisco.

With the election of Governor Gray Davis in 1998 a new group of gubernatorial appointees joined the two assembly and two state senate appointees on the High-Speed Rail Authority Board. All were then enthusiastic supporters of high-speed rail. Governor Davis declared his intent to have high-speed rail as his legacy project and encouraged the election of new appointee Rod Diridon Sr. as the chair, which occurred in early 2001. With that mandate and more favorable annual budgeting, the project moved to the formal Program-Level Environmental Clearance (Program EIR) phase, the required next step in the federal environmental clearance process. The selection by competitive bid of the five corridor segment project leaders and consulting teams catapulted the project into high gear. The small executive staff under Morshed and project management team under Parsons Brinckerhoff's Anthony Daniels scrambled to keep up but had no authorization to add badly needed oversight staff. As some of the corridor elements seemed to be falling behind schedule, Diridon declared emphatically that any consulting team that couldn't maintain the ambitious schedule would be replaced, which did indeed occur with one team leader.

The Program EIR was tasked with examining every corridor, train technology, and station location alternative. Those alternatives were narrowed to select one combination (corridor, train technology, and set of station locations) that offered the best economic and operational characteristics while creating the least environmental impact. The project was massive, but the expectation was to complete the Program EIR in approximately four years. Unfortunately, the collapse of the "dot-com" boom hit the state hard after 2002, followed by serious energy shortages that resulted in a successful recall of Governor Davis and the election in 2003 of Governor Arnold Schwarzenegger (California High Speed Rail Authority, 2015).

HIGH-SPEED RAIL BOND APPROVAL

High-speed rail continued to be popular with California voters up to and during this period of transition. On September 19, 2002, as State Senator Jim Costa prepared to retire, his bill (SB 1856) authorizing a $9.95 billion general obligation HSR bond to be placed on the ballot was approved by the legislature and signed into law by Governor Davis. The bill-signing ceremony at the California Railroad Museum in Sacramento was heavily attended, with the State's leadership anticipating significant progress on the project.

The intent was to take the Costa HSR Bond measure to the voters in 2004, but that ballot was encumbered by competing tax measures relating to economic and energy crisis recovery. The 2006 ballot was similarly encumbered, so the high-speed rail bonds were delayed once again. Finally, then-Merced Assembly member and now-State Senator Kathleen Galgiani, who chaired the legislature's high-speed rail coalition, carried an adjustment to the previously approved language and placed "Proposition 1A" on the 2008 ballot. With enthusiastic election support by the construction, labor, environmental, manufacturing, and commerce communities, the measure passed in November of 2008 with 56 percent of the vote statewide.

Note: Unfortunately, the cost of the project identified in the 2002 concept study had ballooned from approximately $25 billion to $46 billion during the more detailed Program EIR, and then to $60+ billion when the specific community requirements (underground and elevated structures, etc.) were identified. The costs were then inflated to construction-year (2025 to 2030) dollar values, later reaching $98 billion as noted in 2012 using even more conservative business plan studies. Through value engineering the project was refocused on the planned original operating segment from Anaheim via Los Angeles and the Central Valley to San Jose and San Francisco. That segment of approximately 540 miles was projected to cost $68 billion in construction-year dollars. As a result, the $9.95 billion in state bond revenue, which was to cover 40 percent of the project, became a significantly smaller portion of the final project's capital budget. Consequently, Governor Schwarzenegger's public-private partnership concept was more urgently needed and began to be carefully planned.

NEW LEADERSHIP, NEW CHALLENGES

In his campaign for governor, Arnold Schwarzenegger tepidly declared his support for high-speed rail and, once elected, did continue to offer assistance. Though he retained Diridon, he replaced his other four CHSRA Board appointees with more conservative members committed to seeking a public-private partnership to build the project. Former Republican Assembly Speaker Kurt Pringle, then the Mayor of Anaheim and a high-speed rail supporter, was elected in 2006 to chair the Board and was joined by David Crane, a San Francisco venture capitalist and financial advisor to the governor. Led by Pringle, who was determined to have a major HSR terminal in Anaheim, the new more conservative yet still supportive board majority began functioning but with the intent to change the staff and consulting leadership. That occurred in 2007 with the "retirement" of the long-time, venerated Executive Director Mehdi Morshed and the change of Tony Daniel as the Parsons Brinkerhoff project manager.

After an international executive search, respected rail project manager Roelof van Ark was hired as the new CEO with the approval of the new board majority.

Though an outstanding engineering manager, the new CEO lacked experience with the National Environmental Protection Act (NEPA) and the California Environmental Quality Acts (CEQA) which invite in-depth community involvement with all aspects of the planning and environmental certification process. He was pressed by the governor and CHSRA board to accelerate the project and was unsympathetic to delays or criticism. An example was his less-than-cordial testimony before the State Senate Transportation Subcommittee of the Budget Committee, chaired by Palo Alto Senator Joe Simitian. Three small but affluent communities (Palo Alto, Menlo Park and Atherton) objected to the idea of high-speed rail facilities in their towns. Objections were also voiced by some key constituents of the state's political community, including the Farm Bureau and anti-tax and property rights groups, among others. The CHSRA outreach effort was not as extensive, forthcoming or consistent as the situation required. Thus, the expected legal and political challenges began with a vengeance.

ENVIRONMENTAL CERTIFICATION AND BID PREPARATION

During the 2008 to 2010 period the final Program EIR identified the preferred route, station sites, and operating technology (220 mph, electric, steel-wheel technology was selected). In addition, certification for segments of the starter project between San Jose and Los Angeles began to fall into place. The Authority's efforts began to shift to the Project-Level Environmental Review (Project EIR). The alternatives (elevated, at-grade, trench, or tunnel) for every segment of the designated corridor were evaluated and a preferred alternative for each corridor element was to be recommended. Once legally certified, the final corridor segments were to be engineered to the 15 percent level in preparation for one or more "design, build, operate, maintain and finance" (DBOM-F) bids.

FEDERAL FUNDING AND STARTER LINE

That pragmatic plan was partially set aside in 2009 when President Obama declared adamant support for high-speed rail during his campaign and in a broadly publicized press briefing at the White House. The subsequent American Recovery and Reinvestment Act and the following year's budget created $10.4 billion in economic stimulus funding for the 11 (later 13) federally designated high-speed rail corridors. Because California was so far advanced on the required studies and had voter support for bond funding, $3.3 billion of those federal funds were granted to the CHSRA project.

A provision in the grant required that construction begin quickly and that an "operating segment" be completed by 2018. An operating segment was defined as a segment of track connecting two substantial metropolitan areas that could achieve independent utility and be put to use immediately. Additionally, the first segment had to be located adjacent to the major rolling stock maintenance and dispatching center, and the track had to be long and straight enough to test trains at an operating speed of 220 miles per hour. The only location fitting those specifications was in the Central Valley between Merced and Bakersfield, a choice the Federal Railroad Administration subsequently accepted.

While some criticized the location of the first segment as unhelpful to the state's crowded urban areas, it has proven to be truly fortuitous. First, the Central Valley is one of the fastest growing regions in the nation with a population of over one million in the Fresno area and nearly 500,000 in the areas surrounding Bakersfield. If the Central Valley were a state it would be the twelfth largest in the union. It would also be the poorest and rank among those with the worst air quality. The region had nearly 30 percent construction industry unemployment and desperately needed the 24,000 jobs per year the project will create for the next five or six years. In addition, the only practical location for the maintenance base is in the center of the system, near Fresno. And the longest stretch of straight track for testing trains is also on this section. More subtly, beginning the project in the middle created healthy competition between the northern and the southern parts of the state to cooperate fully and be chosen for the first extension.

CONTRACTING AND SCHEDULE

Once the Merced-to-Bakersfield starter section is completed by 2018, a determination will be made by the CHSRA Board regarding the type of interim service that will operate in the corridor until the connection to Palmdale can be completed. California HSR will then feed into the Los Angeles Metro Link commuter train at Palmdale. The Capital Corridor and the Altamont Corridor Express intercity lines north of Merced will connect to Sacramento and the Bay Area until the high-speed rail connection via the Pacheco Pass to the Bay Area is completed in the mid-2020s. Thus, Los Angeles and San Jose will be connected to the Central Valley segment by currently operational commuter and intercity rail systems until the 2020s when their own high-speed rail segments are completed.

For the Merced-to-Bakersfield starter element, the planners have used design-build bidding (DB) instead of DBOM-F. Five international bid teams have been prequalified to compete. The bid for the first $1.1 billion civil work segment from Madera to Fresno was won by Tutor/Perini/Zachary/Parsons, a joint venture. Preliminary construction work is in progress, as is the acquisition of the remaining right-of-way. It is important to note that Construction Package 1 was let at over $100 million under the engineer's estimates, providing significant cost savings. The next sequence of two DB bids, extending south of Fresno, was advertised in late 2014 with three of the five prequalified bidders responding. That combined construction Packages 2 and 3 is for the civil work between Fresno and Bakersfield. Joint venture Dragados/Shimmick/Flatiron (DSF) won that bid and will begin the work on the Central Valley corridor in early 2015. The fourth and fifth Central Valley bids will be for the remaining civil work along with the rails and signals for the full 140-mile Central Valley segment. The second contract was let at more than $400 million below the engineers' estimate. The total savings for the first two contracts is over $500 million below the estimated cost. If that trend continues, enough will be saved to begin the extension south of Bakersfield toward Palmdale using currently available funds.

While the starter segment is progressing, the rest of the system continues to be planned for one or more design-build bids, with public-private partnership funds now expected to cover approximately half of the project costs. The strongly expressed international interest suggests that major rolling stock manufacturers from around the world, backed by countries and/or government-owned operations, may organize extensive bid teams. Those comprehensive groups may include design, construction, rolling stock manufacturing, system operation and maintenance, finance, bonding, insurance, and more. In return for designing (to preapproved specifications), building, operating, maintaining, financing, and paying a competitive annual franchise fee, the winning bidder could obtain an exclusive franchise for 30 to 50 years, receiving the profits from the system during that period. An additional attractive component of this bidding process is that the winner will have the first true high-speed rail train manufacturing capacity in North America. That capability creates a major competitive edge as the other 12 approved US high-speed rail corridors eventually proceed into development.

LEADERSHIP AND CONSISTENT FUNDING

Transformational infrastructure projects require strong political leadership over time. For the California high-speed rail project that bipartisan leadership over the span of four governor's terms dating to the mid-1990s has been the difference between success and failure. Governors Wilson (R), Davis (D), and Schwarzenegger (R) have sustained the project during the toughest of economic times. Governor Edmund G. "Jerry" Brown's (D) courageous, sincere, and tactically brilliant leadership has enabled the project to make the most significant progress of any high speed rail project in the nation's history. That strong forward momentum brought on by public investment and consistent policy leadership is attracting the private investments needed to complete the system.

Governor Brown's determination to make the California High-Speed Rail Authority a world-class organization began with the appointment of current Authority Chair Dan Richard and the selection of former Caltrans Director Jeff Morales as CEO. Under their leadership, the Authority's staff has grown from under a dozen to nearly 175 staff members. The growing team has taken on much of the responsibility for the development of the statewide project, and the fine Parsons Brinckerhoff team has moved to more of a supporting role. Having also achieved a full complement of Board members, the Authority is now poised to deliver on the promise of the high-speed rail system California voters envisioned when the bonds for the project were approved in 2008 (California High Speed Rail Authority, 2015).

At the request of Governor Brown, for the 2014 budget the progressive California legislature approved the permanent dedication of 25 percent of the fees collected, from the newly created cap-and-trade program, to the high-speed rail project. That new program encourages the reduction of greenhouse gas emissions (GHG) by California industry by levying fees on the most significant polluters. Those fees are then invested in GHG-reducing strategies. When bonded over the next 30 years HSR's 25 percent will generate between $10 and $15 billion. Combined with the federal grant and state bond funds, the resulting nearly $30 billion will allow completion of the project from Merced through Fresno to Bakersfield and Palmdale, and add the "bookends" between Burbank, Los Angeles' Union Station, and Anaheim in the south, and San Jose to San Francisco in the north. The bookend projects will include an upgrade to high-speed rail standards, including electrification and positive train controls, of the Metro Link and Caltrain systems to safely allow, joint, integrated use with high-speed rail.

While securing an ongoing source of funding is critical to the continued viability of the project, the cap-and-trade investment may have an even more important side-effect. Seeking private investment has been a necessary part of the program's overall strategy for full funding. However, as large infrastructure projects around the world have shown, private investment comes forward only after significant public investment has been achieved. Cap-and-trade funds should serve as a catalyst for private investment to come forward because the state government has finally, meaningfully, committed to a long-term investment, providing potential investors with a modicum of security that heretofore did not exist. Since cap-and-trade funds were committed, over a dozen international firms have expressed an interest in developing portions of the project.

THE BLENDED BOOKENDS

Along with the influx of new funding, it is important to point out that notable components of the California High-Speed Rail project have already begun, namely, the construction of the San Francisco Transbay Terminal Center, the Anaheim Regional Transit Intermodal Center, and the Caltrain electrification project. In April 2012, the California High-Speed Rail Authority approved the 2012 Business Plan, which laid out the phasing and funding plan going forward. In addition, the business plan called for the development of a "blended" system in the highly urbanized areas of Northern California (San Francisco to San Jose) and Southern California (Burbank to Los Angeles to Anaheim). In the Bay Area, Caltrain's Peninsula commuter corridor proved to be too narrow for a fully built-out, four-track system with separate high-speed tracks – a plan that also encountered significant community opposition. The revised 2012 Business Plan introduced the concept of a blended rail system where Caltrain commuter trains and high-speed trains would operate on the same rails protected by positive train controls. The first step in that direction is the full electrification of the Caltrain rail corridor between the Fourth and King Streets Station in San Francisco and the Tamien Station in San Jose. The state legislature authorized $600 million in Proposition 1A funds toward the electrification project. This investment, along with funding from regional partners, including the counties of San Francisco, San Mateo, and Santa Clara; the cities of San Francisco and San Jose; and the Metropolitan Transportation Commission and BART, earmarks $1.5 billion to fully electrify the Caltrain corridor.

Titled the Caltrain Modernization Program, the electrification of the Caltrain corridor lays the foundation for future high-speed rail service in the Bay Area. Caltrain expects to complete environmental clearance by early of 2015. A design-build contract will be put out for bid, and construction is expected to begin in late 2016. Electrically powered revenue commuter service of six trains per hour per direction is expected by 2020, replacing the current well-used diesel push-pull system. The High-Speed Rail Authority will follow with an environmental review in mid-2015 for blended service introducing up to four high-speed trains per hour per direction soon after the environmental clearance for electrification in the corridor is certified. A similar process is planned for the Metro Link commuter rail corridor from Palmdale to Anaheim in the Los Angeles area.

REGIONAL TERMINAL PLANNING AND CONSTRUCTION

A video on the San Francisco Transbay Joint Powers Authority website describes the history of the Transbay Terminal. The narrator explains that before World War

II over 24 million passengers annually traveled through the Transbay Terminal by rail. The old Key Rail System operated on the lower deck of the Bay Bridge, and traveling by regional rail was the order of the day. Twenty-four million rail passengers annually! Sadly, after the war, the auto became king, and the rails were pulled up from the bridge and elsewhere. The region is now working to reestablish the same service at hundreds of times the cost. The new Transbay Terminal Center is under construction and rising out of the rubble of the old terminal. The objective is to provide a new generation of multimodal transportation services for future generations of Californians. The new Transbay Terminal will be among the largest multimodal transportation facilities in the country, providing convenient connections to regional rail, rapid transit, and local and interregional bus and light rail systems, all from the heart of downtown San Francisco. The Transbay Terminal is expected to be completed by 2018.

San Jose is planning similar, though not quite as expansive improvements for San Jose Diridon Station serving Silicon Valley. Feeding the station is Caltrain (expanded in the future to include feeders from Monterey and Santa Cruz), the Altamont Corridor Express, Capital Corridor service, 50 miles of regional Valley Transportation Authority light rail, numerous express and local route bus lines, shuttles and other amenities. The BART extension to San Jose is currently under construction. An additional 50+ route-miles of light rail and a proposed automated guideway airport shuttle may be added before high-speed rail is built. San Jose's Diridon Station is expected to host over 800 trains a day when high-speed rail arrives in the late 2020s.

Electrification of the Caltrain Corridor and the construction of the Transbay Terminal project can be considered among the first high-speed rail projects in California but comprise just a small portion of the statewide project. The influx of stable funding and momentum from the start of construction projects in the Central Valley and the Bay Area allow for the advancement of project sections in both Northern and Southern California. Those projects, dormant for many years, are now proceeding, creating current electrically powered commuter rail mobility and future high-speed rail options at a greatly reduced cost.

LEGAL ACTION FAVORS CHSRA

Large infrastructure projects, especially those affected by the California Environmental Quality Act, attract lawsuits. The 18-year effort to design and engineer California's high-speed rail system has seen the expected deluge of legal actions. Some are merited, many are frivolous, and all have the potential to cause expensive delays.

For perspective, however, it is important to note that a project no less venerable than the Golden Gate Bridge was challenged by more than 2,300 lawsuits filed by the time the project commenced construction in 1934. Each was won or settled by the time the bridge was completed in 1937. Who today could imagine the Bay Area functioning without that iconic piece of the nation's transportation infrastructure? The dozen or so lawsuits brought against the CHSRA to date have either been settled or ruled in favor of the project, and, thus far, the State Supreme Court has declined to hear appeals. The courts have declined to take any action that invalidates current contracts or impede future contracting and the sale of the HSR bonds has been affirmed. Recently the federal Surface Transportation Board has assumed authority over the project thereby eliminating the legal requirement for future adherence to CEQA though NEPA continues to apply. Though there will be future legal actions none currently have serious potential to further delay or stop the project.

FORMAL GROUNDBREAKING

The symbolic, very successful, formal groundbreaking for the project occurred before over 1,000 rail fans and appreciative citizens in Fresno on January 6, 2015. After a welcome by Fresno Mayor Ashley Swearengin, CHSRA Board Chair Dan Richard offered a historic overview, shared the planned completion schedule, and guided the day's program. The event featured upbeat comments by Governor Jerry Brown, Congressmember Jim Costa, US EPA Administrator Gina McCarthy, Federal Railroad Administrator Joseph Szabo, and others. Several of the presenters singled out Richard and CEO Jeff Morales for reinvigorating the project in 2010 at the beginning of the governor's third term. The 76-year-old governor, just beginning an unprecedented fourth term, stole the show with an extemporaneous, environmentally focused speech. To great applause, he said he intended to eat his vegetables and exercise so he would be around in 2030 at 92 years of age to ride the first through train from Los Angeles to San Francisco. Also introduced were Chief of Staff Nancy McFadden; California Secretary of Transportation Brian Kelly; Caltrans Director Malcolm Daugherty; and other officeholders, including current CHSRA members Catherine Estolano-Perez, Dan Richard, Tom Richards, Mike Rossi, Lynn Schenk, and Thea Selby; past CHSRA chairs Diridon and Pringle; and many others. Lynn Schenk, long-time CHSRA Board member, former Congressmember, state cabinet member, and chief of staff to Governor Davis, was lauded as the "mother" of high-speed rail. The celebration ended with the delegation signing a rail that will be permanently displayed in the California Railroad Museum in Sacramento (Figure 1).

Figure 1.
Source: California High Speed Rail, 2015, http://www.hsr.ca.gov/

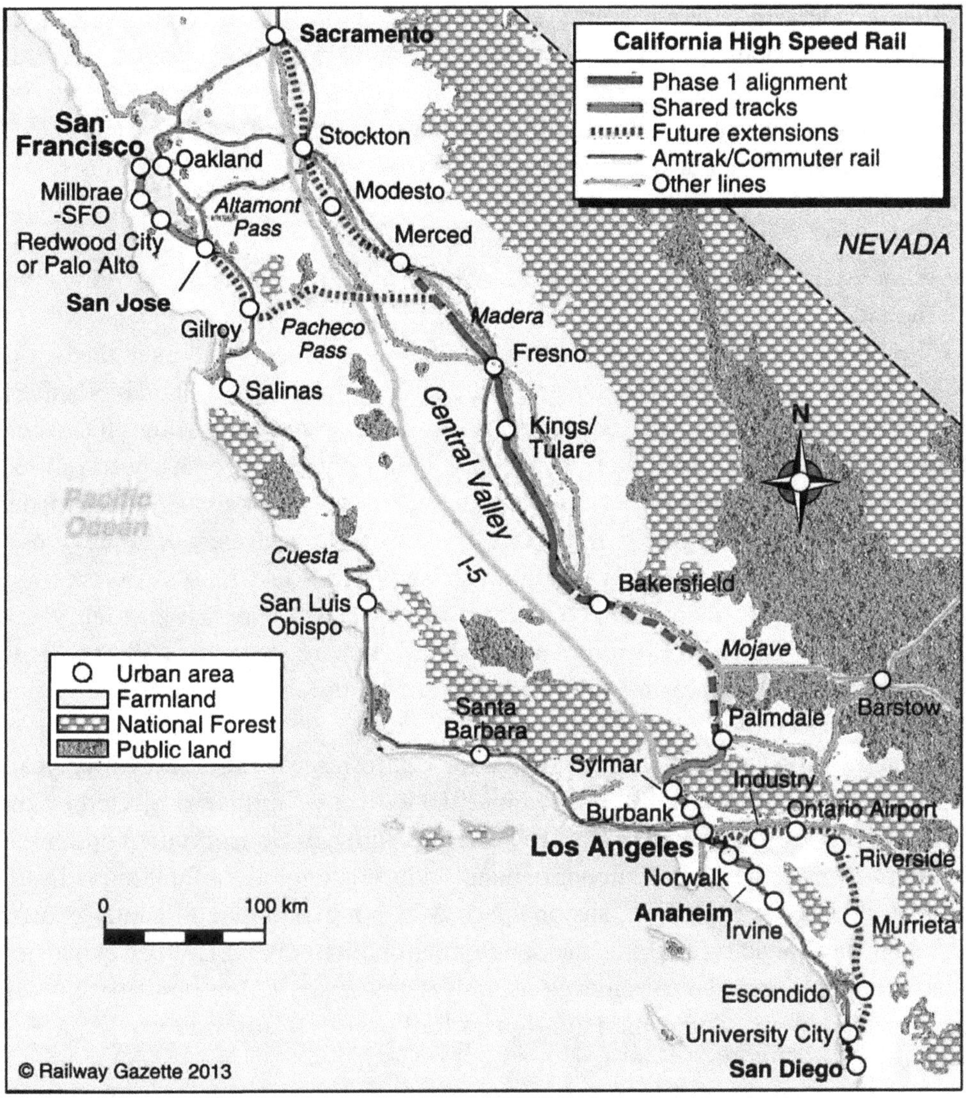

CONCLUSION

The California high-speed rail project has achieved a critical mass not yet seen for HSR in North America. But is high-speed rail right for California? The question is not only can we do it, but should we? The authors of this chapter believe the answer is a resounding yes.

First, traffic congestion continues to be a growing problem in California. Six of the thirty most congested areas in the country are located in California. The Silicon Valley Leadership Group recently released a survey of its member CEOs – heads of the top 400 companies in the Bay Area (among them, Apple, Google, HP, Intel, Yahoo, and others). These corporate leaders rated traffic congestion among the five most challenging issues in maintaining a business and promoting economic expansion in the Bay Area. Further, three of the five most heavily travelled Amtrak routes are in California (Capitol Corridor, Coast Starlight, and the San Joaquin Regional Rail System). Californians have been compelled to seek new travel modes to avoid ever-increasing roadway congestion, and California is recreating a rail-riding ethic in the process.

Second, California has one of the most congested airline corridors in the world. A significant percentage of flights originating at Bay Area airports land in Southern California. Comparably, in a recent review of the literature by the Mineta Transportation Institute at San Jose State University, in regions where high-speed rail has been introduced, there has been significant shift away from conventional modes of travel, especially air travel. In the Northeast Corridor, travelers who previously flew between New York and Washington are switching to ACELA's Amtrak train, boosting its market share from 37 percent to 75 percent over the past ten years. Between New York and Boston, Amtrak's market share went from 20 percent to over 54 percent. Downtown-to-downtown on trains instead of planes is becoming a mantra, especially for business travelers.

Finally, projected population growth for California over the next thirty years continues to be explosive (38 million in 2010 to over 60 million projected by the 2060s). This growing population must be served with new transportation options as traditional modes are buried under demand-induced congestion. Funding is insufficient, and highway lane expansion space does not exist to significantly expand the highway capacity entering the state's metropolitan areas. Even if expansion space did exist, in order to achieve the equivalent transportation capacity that will be provided by the California High-Speed Rail project, California would have to construct over 4,300 miles of new roadways, 115 new airline gates, and at least 4 new airport runways. These improvements would cost two to three times the current $68 billion price tag for high-speed rail without any benefit to the environment and would have to be repeated at mid-century as capacity would again be exceeded by demand. Note that rail system capacity can be doubled or tripled by adding cars to a train or trains to the tracks. Highway capacity can be significantly expanded only through the expensive and disruptive process of adding lanes.

In short, the cost of meeting the burgeoning trip demand by building more highways and airports would cost in excess of $250 billion while to build the high speed rail and feeder systems would cost less than one half that amount. And the highways and airports would be beyond capacity again in less than 50 years requiring another huge investment of funds to meet the state's trip demands of the second half of the 21st century. High speed rail and the feeder lines can double and triple capacity by inexpensively adding trains to the tracks instead of building expensive additional right-of-ways. So the horrendous cost of temporarily meeting the burgeoning population's demand for highway and airport capacity would, at a fraction of that cost, be instead met for the next century by building sustainable transit systems. And the lynch-pin of that long-term, less expensive and visionary remedy is the clean, green electrically-powered high speed rail system.

It is important to note that while high-speed rail is the focal point of the California investment, the overall effort is statewide rail modernization. The HSR project is vital to the state's future mobility but cannot reach optimum potential without improvements to the regional rail feeder systems that connect to the high-speed backbone.

In 1963, President John F. Kennedy spoke before the Irish Parliament of the importance of looking to the future. He said of Ireland, "This is an extraordinary country. George Bernard Shaw, speaking as an Irishman, summed up an approach to life: Other people, he said, 'see things as they are and … say "Why?"... But I dream things that never were – and I say: "Why not?"' It is that quality of the Irish – that remarkable combination of hope, confidence, and imagination – that is needed more than ever today."

It is that same quality that Californians espouse and that is exemplified by the development of the California High-Speed Rail project.

Kennedy went on to say, "The problems of the world cannot possibly be solved by skeptics or cynics, whose horizons are limited by the obvious realities. We need men [and women] who can dream of things that never were and ask, why not?" As Governor Brown has said, "It is time for California to do great things again." We built the Golden Gate Bridge and the Bay Bridge during the deepest darkest depression in our history and recently rebuilt the eastern span of the Bay Bridge during the worst recession in our history. Our parents and grandparents built the University of California system, the California State University system, and the State Water Project – improvements that have made California the world leader in technology, agriculture, and public policy, with an economy that leads the nation in fiscal responsibility.

In the end, high-speed rail is about choosing our future and having the confidence in ourselves to make that future as wonderful as we choose. California has a history of building legacy projects, and we have an obligation to leave a better state than the one we inherited, despite the challenges ahead. We are living off an infrastructure built five generations ago. High-speed rail is one of many changes we need to make if we are to remain the eighth-largest economy in the world.

It is time for our generation to continue that tradition and establish California's legacy for the next hundred years.

REFERENCES

California High-Speed Rail Authority. (n. d.) Retrieved from http://www.hsr.ca.gov/

Mineta Transportation Institute. (n. d.). Retrieved from http://transweb.sjsu.edu/

Section 2

Chapter 3
Comparing the Size of the Market in Potential High Speed Rail Corridors of the Midwest:
A Gravity Model Approach

Joseph P. Schwieterman
DePaul University, USA

ABSTRACT

Transportation planners in the Midwest are advancing an agenda for high-speed rail while simultaneously investing in numerous corridors that have different strengths and weaknesses. To assist in establishing priorities, this paper compares the size of the travel markets in seven of these Midwestern corridors. Using a technique known as the gravity model, it demonstrates that the corridors linking Chicago to Cincinnati and Detroit likely exceed the others on the basis of total passenger traffic, and, along with the Chicago – Twin Cities route, rank at the top with respect to the passenger-miles of travel. The routes to Cleveland, St. Louis, and other locales compare less favorably. The top corridors, while shown to be only a fraction of the size of the California and Northeast Corridor routes, are found to be appreciably larger than most of the country's proposed high speed rail corridors, suggesting they have considerable potential for rail-passenger development.

DOI: 10.4018/978-1-5225-0102-2.ch003

INTRODUCTION

A fundamental issue affecting the development of high-speed rail service throughout the Midwestern states is a dispersion of efforts across numerous corridors. Resources have been spread out over a variety of routes on account of political boundaries, differing goals of government agencies, challenges of each corridor, and a lack of consensus on which route deserves to be the top priority. Although lengthy segments of the Chicago – St. Louis and Chicago – Detroit routes have been upgraded for 110 mph service—the standard used in this chapter for a service to be considered "high speed"—it remains unclear when these routes will become fully functioning high-speed corridors, due to remaining obstacles.

This dispersal of resources across various corridors has not been nearly as significant in the development of high-speed service in most other parts of the world. In France and Japan, for example, a consensus emerged early in the planning process that the Paris – Lyon and Tokyo – Osaka routes, respectively, were the highest priorities. In Spain, resources initially focused on completing the Madrid– Seville route before significant resources were devoted to other routes. A similar consensus has emerged about which corridors in California and the Northeast states are most suitable for new or improved high-speed trains. Due to the reasons listed above, such a consensus does not exist in the Midwest.

This study seeks to evaluate the size of corridors potentially served by the Midwest Regional Rail System (MWRRS), the multi-state system envisioned to promote faster and more frequent rail service in the region, using a technique known as the "gravity model." This approach considers the populations of the metropolitan regions served by various corridors, the distances between them, and the rate at which the volume of travel diminishes as distances rise. This gravity model approach is based loosely on the Newtonian law of gravity, which dictates that any two bodies attract each other with a force that is directly proportional to the product of their masses and inversely proportional to the square of the distance between them.

This report's primary goal is to provide a fresh perspective on the traffic potential of Midwestern corridors due to the growing obsolescence of past efforts to compare them as well as foster greater understanding of the tradeoffs the region faces. In the following sections, the analysis explores seven corridors, originating from Chicago, including: Chicago – Cincinnati, Cleveland, Detroit, Lansing/Port Huron, St. Louis, Minneapolis/St. Paul, and Omaha. Results from the gravity model are compared with estimates used in a 2004 study of the MWRRS. The final section compares Midwestern corridors with those in other parts of the United States, including the Cascade, California, and Northeast corridors.

BACKGROUND

This study is not the first to evaluate the merits of various Midwestern HSR corridors. Several research publications offer a systematic evaluation of the strengths and weakness of various routes proposed for the region. Hagler and Todorovich (2014) use an "index approach" that considers many factors, including airport and highway congestion to rank the importance of prospective HSR corridors across the country. Their analysis ranks the Chicago – Detroit and St. Louis corridors as the highest priorities in the Midwest.

Most studies advocate for the system as a whole rather than for specific corridors. The Midwest High Speed Rail Association has done extensive analyses demonstrating the economic value of the various corridors under consideration (2011). The Federal Railway Administration has designated corridors that meet certain criteria as being eligible for high-speed rail funding, but stops short of attempting to prioritize these routes (2014). Few direct comparisons between routes, consequently, have been made.

A study commissioned for the MWRRS in 2000 compares the capital costs and implementation issues affecting each corridor (TEMS, 2004). Chapter 4 of the study includes estimates of passenger travel by mode in each corridor encompassed by the system. The study, while comprehensive and the product of extensive statistical analysis, relies on data almost entirely from the 1990s to measure vehicular traffic and shies away from directly comparing the corridors according to their passenger-carrying potential. As noted later in this report, the authors of the study are candid about the measurement problems that affect their assessment of each corridor.

The lack of up-to-date estimates on travel activity in Midwestern corridors can be blamed in part on the federal government's discontinuance of the American Travel Survey after it was administered in 1995. As a result, and which may come as a surprise to some readers, no reliable data presently exists on total travel between metropolitan areas. Although traffic counts are available for Amtrak and air travel, these modes account for only a small share of the total traffic in corridors presently being considered for improved rail service. All recent estimates of the size of Midwestern markets, consequently, are somewhat speculative in nature.

Apart from the problem of drawing upon data sets that are subject to considerable error, most existing studies on Midwestern corridors measure travel behavior under sharply different market conditions that prevail today. Almost all were made prior to the spike in fuel prices that occurred in 2005, and most were conducted before the low-cost intercity bus service became available in the region (first by Megabus and later by other carriers) in 2006. Almost all were made prior to the start of the economic downturn that began in 2008.

APPROACH

The approach used in this study, while relatively simple, provides a useful starting point in helping the region establish clearer priorities. The gravity model is based on the following formula:

$$\text{Passenger traffic} = \frac{k\left(P_1 P_2\right)}{d^b}$$

P_1: Population of the first city
P_2: Population of the second city
d: The distance separating them
k: A constant
b: A parameter measuring the rate of transport "friction" associated with travel between two locations

The variable *b*, or "friction," reflects the rate at which travel demand diminishes as distance increases. This will depend on the cost, speed, and hassle associated with travel (Haynes and Fotheringham, 1984.) A decline in fuel costs, for example, would cause *b* to decrease, while rising fuel costs, or additional tolls, would cause *b* to rise. More sophisticated models, of course, consider many other variables, such as socio-economic, demographic, and geographic factors, while still drawing upon this fundamental size/distance relationship. Even so, as shown below, using the simple model described is a valuable exercise that can provide new insights and point the way for additional research.

The concept behind the gravity model is illustrated visually with three population centers represented by their size in *Figure 1*. If all other things were equal, one would expect travel in the first city pair, between P_1 and P_2, to be smaller than between P_2 and P_3 due to the fact that:

- Population of P_1 is smaller than P_3, and
- The distance associated with d_2 is greater than d_1.

The extent to which travel differs in these combinations of cities will depend on the rate of friction (*b*). The higher the friction, the greater the extent to which $P_2 - P_3$ travel will exceed that of $P_1 - P_2$.

Perhaps the greatest weakness of using this approach to compare corridors is that it utilizes only the second step of what is traditionally a two-step travel estimation process. (See discussion in endnote section.) Similarly, the model ignores the effects

Figure 1. Illustration of the gravity model

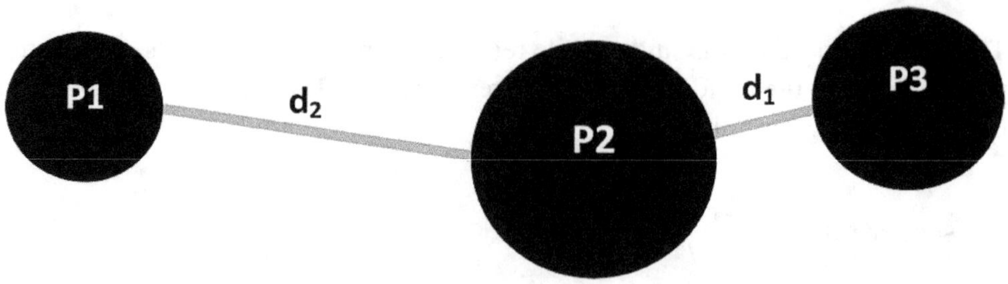

of crossing state lines, as well as income levels, population growth, and affinity factors that will affect travel, such as the presence of colleges and universities, patterns of commerce, and the ethnic settlement patterns of years ago.

The simplicity of this approach nonetheless offers three important advantages: i) it avoids mixing data from different sources, which can introduce biases related to timing issues and differences in the way information is collected; ii) it allows for comparisons without the undisclosed adjustments that have been known to generate skepticism by many observers, some of whom perceive the models as a "black box" able to generate whatever results the clients seek to show; and iii) the method can be easily explained in a way that even stakeholders without a strong background in statistics can understand. At a minimum, the gravity model approach can be a jumping off point for future analysis and serve as a useful tool to illustrate complex demand-forecasting concepts to policymakers.

MODEL INPUTS

The model is calibrated as follows. The population of metropolitan areas is based on 2012 U.S. Census estimates. Only metropolitan statistical areas (MSAs) with populations of 100,000 or more are considered. Travel distances are based on great-circle (air) mileages rather than surface-travel distances. The population of each MSA, and their respective distance to Chicago, appears in *Table 1*. The arrangement of population centers along the seven corridors is shown on *Figure 2*. This "data map" shows each population center proportional to their 2012 estimated population.

The analysis considers travel in all the combinations of population centers along a corridor. For example, on the Chicago – St. Louis route, it considers travel between the endpoints as well as Chicago – Springfield (197 miles), Bloomington – Springfield (84 miles), and so forth.

Comparing the Size of the Market in Potential High Speed Rail Corridors

Table 1. Metropolitan areas along Midwest corridors: Population 000s and distance to Chicago

	Miles to Chicago	2012 MSA Population	Projected MSA 2030 Population
Chicago –Twin Cities, MN			
Milwaukee, WI	82	1566.98	1770.10
Madison, WI	122	620.77	740.68
La Crosse, WI	226	135.29	163.48
Minneapolis-St. Paul, MN	355	3,422.26	4,397.27
Chicago–Lansing-Port Huron, MI			
Kalamazoo, MI	109	330.03	373.51
Lansing, MI	167	1,005.64	1,211.50
Chicago – Detroit, MI			
Kalamazoo, MI	109	330.03	373.51
Ann Arbor, MI	201	350.94	437.37
Detroit, MI	237	4,292.06	4,888.78
Chicago – Cleveland, OH			
Ft. Wayne, IN	140	421.40	485.03
Toledo, OH	211	608.71	682.41
Cleveland, OH	308	2,063.53	2,201.72
Chicago – Cincinnati, OH			
Lafayette, IN	108	206.41	221.43
Indianapolis, IN	165	1928.98	2173.26
Cincinnati, OH	252	2,128.60	2,471.84
Chicago – St. Louis, MO			
Bloomington-Normal, IL	118	188.71	232.46
Springfield, IL	179	211.99	216.25
St. Louis, MO	263	2,795.79	3,138.59
Chicago – Omaha, NE			
Quad Cities, IA-IL	154	382.63	408.82
Des Moines, IA	309	588.99	653.55
Omaha, NE	435	885.62	1,045.91

Only city pairs involving travel distances of more than 85 miles are considered, a criterion that limits consideration to those *intercity* routes relevant to analysis of HSR. Over shorter distances, travelers can often comfortably drive to the destination in less than 90 minutes, and improvements in train speed (i.e., to 110 mph) typi-

Figure 2. Map of Midwest routes evaluated with metropolitan areas depicted proportionally to 2012 est. population

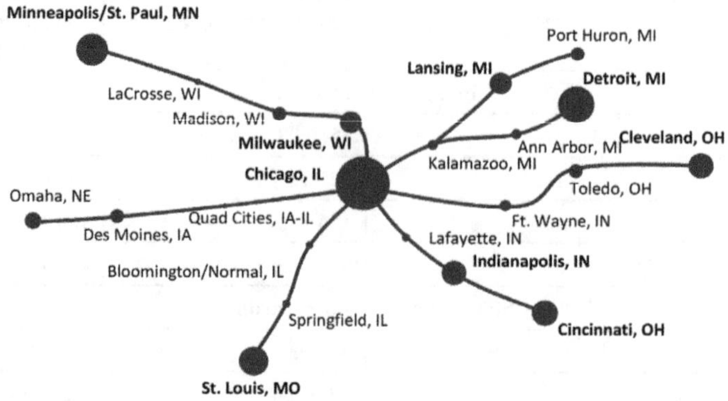

cally reduce travel time only modestly. Rarely, as a result, is the importance of HSR given emphasis in these markets. This criteria results in the exclusion of Ann Arbor – Detroit (36 miles), Chicago – Milwaukee (82), and Kalamazoo – Lansing (60).

There is no universal standard for b (friction). In the scholarly literature, however, b is generally accepted to be between 1.2 and 2.5. The analysis below considers three different assumptions; a moderately low value (1.3), a mid-range value (1.6), and a moderately high value (1.9).

RESULTS

Rather than providing results of the absolute number of passengers moving in a corridor, they are initially presented in *relative terms*. This is due to the fact that determining the absolute number of passengers is not possible without a reliable estimate for k in the gravity model equation, which is unavailable. The results are expressed as a percentage of the Chicago – Cincinnati (via Indianapolis) corridor, which serves as a benchmark for comparison. For example, a corridor with traffic represented as 90% would have nine-tenths of the traffic as the Chicago – Cincinnati corridor.

The analysis suggests that the size of the market is significantly different in the corridors evaluated. The differences depend on whether the unit of analysis is total traffic, passenger-miles of traffic, or the "density" of travel (number of passengers per mile of route).

Total Passenger Travel between Endpoints

This first measure, the amount of travel between the corridor's endpoints, excludes trips to and from intermediate stops. (In this analysis, Lansing is used as the endpoint of the Port Huron corridor on account of the fact the latter community has a metropolitan population of about 40,000, below the threshold for inclusion in the analysis). On this basis, Chicago – Detroit is by far the largest corridor, with 223% of the traffic of the Cincinnati route (which serves as the baseline); the St. Louis route, at 123%, is also well above the average; and the Twin Cities route is near the middle (*Figure 3*). The comparative lack of trips to Cleveland, Lansing, and Omaha are clearly evident.

Figure 3. Passenger trips between endpoints of Midwest corridors as a % of Chicago - Cincinnati route: Moderate "friction" scenario

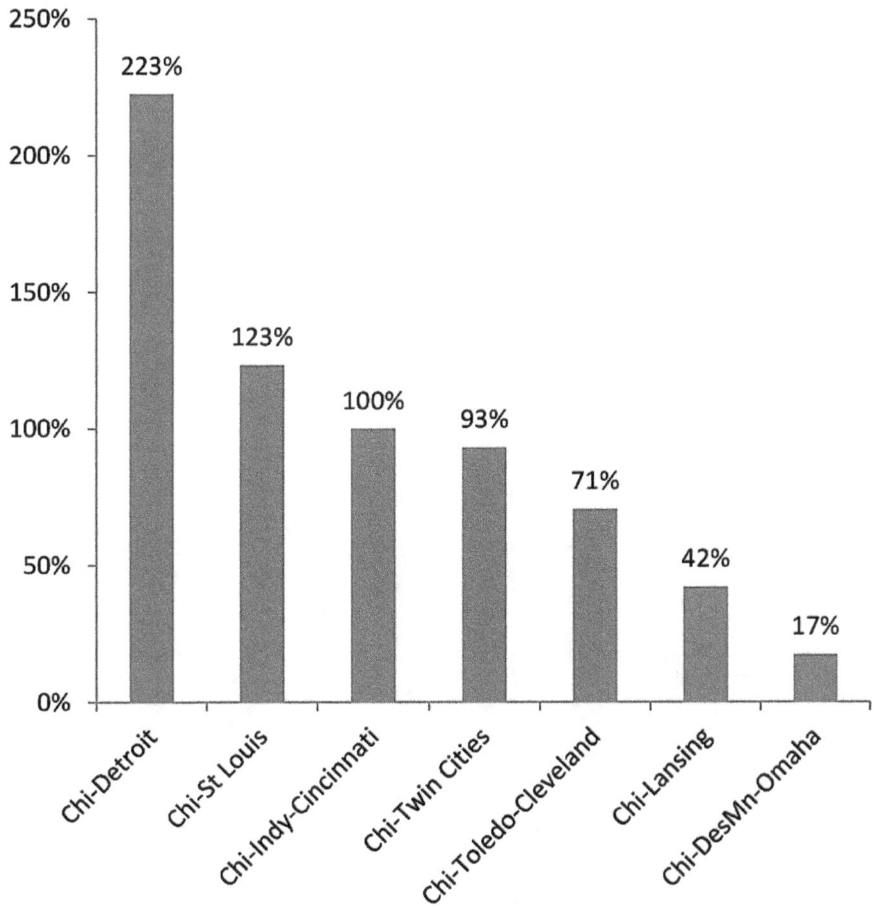

These results illustrate a fundamental issue facing the Twin Cities corridor: although the population of the Minneapolis-St. Paul region far exceeds that of metropolitan St. Louis, the projected amount of travel to Chicago is far less due to the associated travel distance, regardless of the rate of friction considered. At all three levels of friction considered in the analysis, the Detroit and St. Louis corridors easily rank first and second with respect to travel between the endpoints. A summary of the rankings of the corridors under different assumptions appears in the Appendix (*Table 3*).

Total Passenger Traffic Including Intermediate Stops

When intermediate stops are taken into account, and the mid-range scenario for friction is once again used, the Indianapolis/Cincinnati corridor emerges as a top performer (*Figure 4*). The presence of a larger intermediate stop, Indianapolis (pop. 1.9 million) greatly stimulates the volume at both ends of the corridor. Second-ranking Detroit has 80% of this corridor's total passenger traffic. Both routes have traffic volumes that exceed the others by a wide margin.

Figure 4. Passenger travel in Midwest corridors, including intermediate stops as a % of Chicago - Cincinnati route: Moderate friction scenario

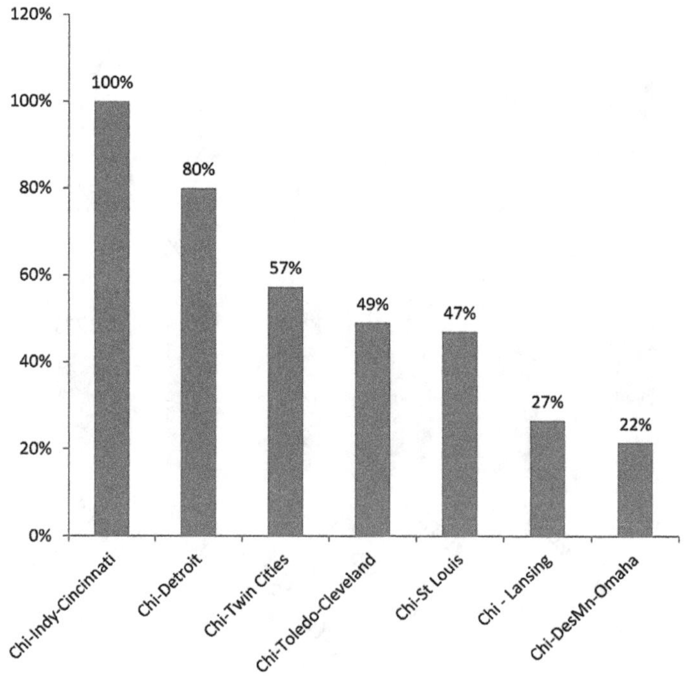

The Twin Cities route's ranking improves using this measure on account of the size of its intermediate cities, particularly Milwaukee and Madison, which are the second and third largest intermediate stops on the Midwestern system, behind only Indianapolis. These intermediate points have a combined population of more than two million. The St. Louis route performs poorly using this measure, due to the limited population of intermediate cities, which have the combined population that is half that of any of the other corridors. (As previously noted, affinity factors not considered in the model, particularly government-related traffic between Chicago and Springfield, IL, may partially compensate for this weakness). When a high-friction scenario is considered, the St. Louis route closes the gap with the Cleveland route. In all three scenarios, however, the Omaha and Port Huron rank at the bottom.

The Cincinnati – Indianapolis leg of the Chicago – Cincinnati corridor is the most densely traveled city pair not involving the Windy City, accounting for more than 20% of that corridor's trips. A word of caution is appropriate, however: these cities may be separated by too short of a distance (99 miles by air and 112 by highway) for trains to be highly competitive with other modes. If the results are adjusted to consider only routes that involve ground distances of about 125 miles or more, the Detroit route is 24% *larger* than the Cincinnati route. The rankings of the other corridors, however, do not change.

Passenger Miles and Density of Travel

The primary objective of transportation systems is to move people across the distance they seek to travel. It is important, therefore, to compare the relative number of *passenger-miles* of travel among the corridors. On this measure, the Cincinnati route is again the largest, but it is essentially deadlocked with the Detroit route, which is 98% as large, followed by the Twin Cities (82% as large). The Cleveland and St. Louis routes are nearly identical on this measure, but rank near the middle, while the Port Huron and Omaha markets are by far the smallest (Table 2).

Table 2. Ranking for Midwestern Corridors by passenger miles of travel as a % of Chicago – Cincinnati Route

1. Chicago-Cincinnati	100%
2. Chicago- Detroit	98%
3. Chicago – Twin Cities	82%
4. Chicago – Cleveland	61%
5. Chicago – St. Louis	60%
6. Chicago – Port Huron	20%

At a lower value of friction (b=1.3), the Detroit corridor marginally outperforms the Cincinnati route, while the Cleveland route falls to a level almost equal to the St. Louis corridor.

Comparing the corridors on the basis of the density of traffic is also instructive. The intensive capital investments necessary for high-speed rail can generally only be justified when large numbers of passengers are carried per mile of route. Such investments cannot be justified when density (passenger miles divided by distance) is low. When evaluated on this basis (*Figure 5*), the Detroit corridor returns to the top and is closely followed by the Cincinnati route. The Twin Cities route is noteworthy for ranking in the upper half, considering that it is more than 25% longer than these top-performing routes (and that Chicago – Milwaukee traffic is excluded).

When viewed in its entirety, the analysis shows the obstacles to justifying high-speed rail developments on lower-ranking routes. At present—perhaps for the reasons describe above—little consideration is presently being given to upgrading most of the Omaha and Port Huron routes beyond that needed for reliable 79 mph service.

COMPARING RESULTS WITH PRIOR ESTIMATES

The above results can be evaluated by considering them against the backdrop of the formal analyses of the Midwest Regional Rail System (Midwest Interstate Passenger Rail Commission, 2000 and 2004.) These studies are richly documented and broad in scope. The consultant that produced them, TEMS Inc., is nonetheless

Figure 5. Density of trips in Midwestern corridors (passenger miles per mile of distance) as a % of Chicago – Cincinnati route

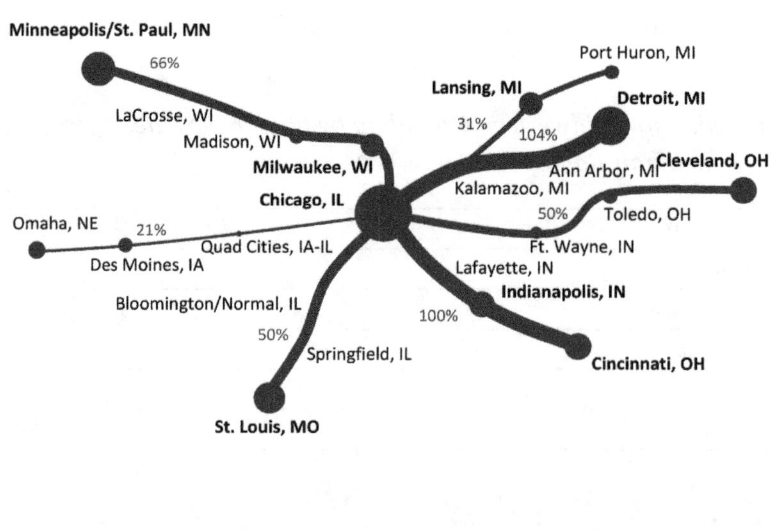

candid about the challenge of drawing up data from different sources, noting, "auto origin destination (O-D) travel data was difficult to obtain and was available only for certain states and regional centers. Estimates on O-D travel for zones that were lacking data were made using the travel characteristics of existing and available data, modified by population, income, employment and trip length."[1] Much of the data used was collected in the early 1990s, with "socioeconomic growth rates" applied to adjust them to levels for the year 2000.

One would expect significant differences between the MWRRS and gravity model results due to the fact that the former looks at *actual volumes* while the latter essentially estimates what travel would be if everything else—including the quality of service—were constant. When considering the demand for a new transportation service, such high speed rail, there are advantages for considering both approaches.

Figure 6 shows the estimated demand for travel in the seven markets using the MWRRS and the gravity model measurements. For the gravity model results, the traffic estimates are made by apportioning the total travel volumes in the seven

Figure 6. Projected travel in Midwest corridors: Estimates annual traffic (in Millions) and gravity model calibrated to MWRRS estimates

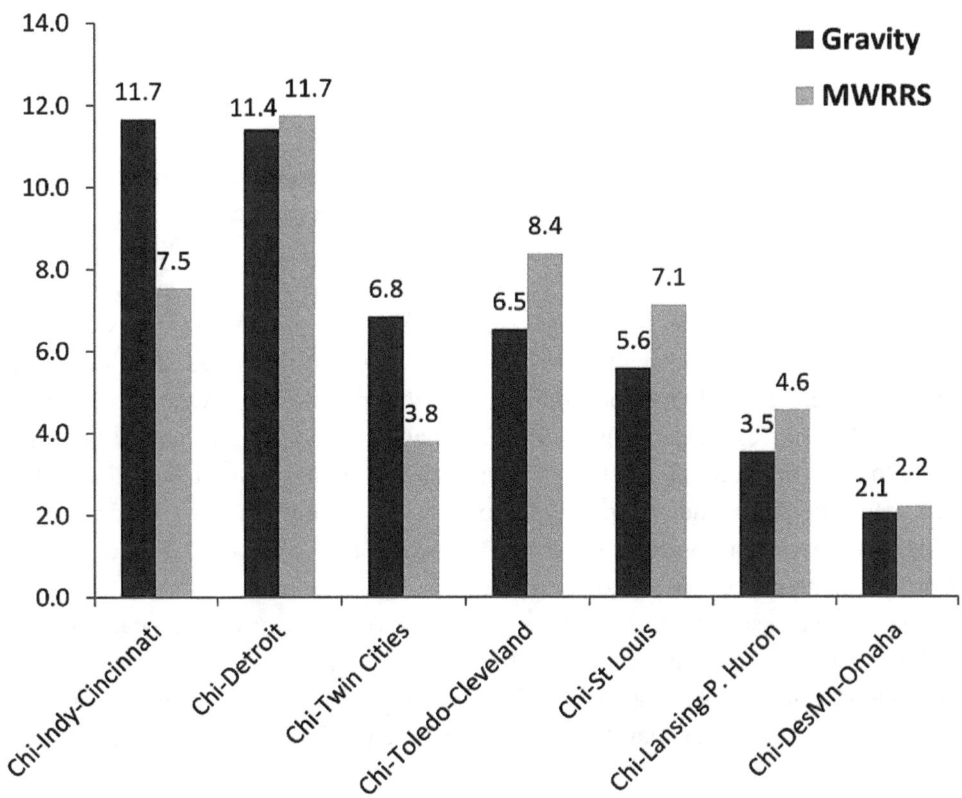

corridors in the MWRRS study on the basis of their estimated relative demand. Moreover, all traffic volumes from this earlier study were adjusted upward by 1.5% annually to transform historical numbers to present-day ones.[2]

There is considerable consistency between the two studies, although some discrepancies stand out. In the gravity model, the Cincinnati corridor is estimated to be much larger and the Twin Cities corridor much smaller in comparison to the MWRRS results. The difference in the first corridor may be partially attributable to the fact that Cincinnati and Indianapolis have both grown faster than the Midwestern average in the decade since the MWRRS study was released. Another factor may be the fact that the gravity model considers Lafayette as an intermediate stop while the latter study does not. The MWRRS estimate may also be pushed downward by usually high cost of airline travel and poor quality of Amtrak service between those communities.[3] There is also reason to believe that the MWRRS study may suffer from a downtown bias due to measurement issues.[4]

On the Twin Cities route, the differences are more difficult to explain. The same factors mentioned in the previous paragraph—changes occurring in the decade since the MWRRS study were released and the absence of low cost air service—may be relevant. Similarly, Amtrak service remains limited to one train daily. Even so, these differences warrant additional analysis.

COMPARISONS WITH OTHER REGIONS

The gravity model offers a practical way to compare the approximate travel volumes among corridors throughout the country. This section compares the Cincinnati and Detroit routes, which were identified as the Midwest's largest, with the Boston – New York, New York – Washington, Miami –Tampa, Houston – Dallas, Portland – Seattle, and Los Angeles –San Francisco routes. As in previous sections, analysis is limited to city pairs in these corridors in which travel distances are 85 miles or more and considers only metropolitan regions with populations exceeding 100,000.

This distance criteria results in the exclusion of notable city pairs, including New York – Philadelphia, Boston – Providence, Orlando – Tampa, and San Jose – San Francisco, from their respective corridors. The comparisons are nonetheless instructive by showing size of the market in which high speed rail would vie with other modes to attract passengers.

The results, based on the mid-range estimate of friction (k=1.6), appear in *Figure 7*. Apart from the Northeast and California corridors, the Midwest corridors are shown to be appreciably larger than all of the others. The Midwestern corridors are at least four times larger than Raleigh – Charlotte rand more than 20% larger than

Figure 7. Passenger volume: Relative to largest Midwest corridor – Trips >85 miles in notable U.S. corridors

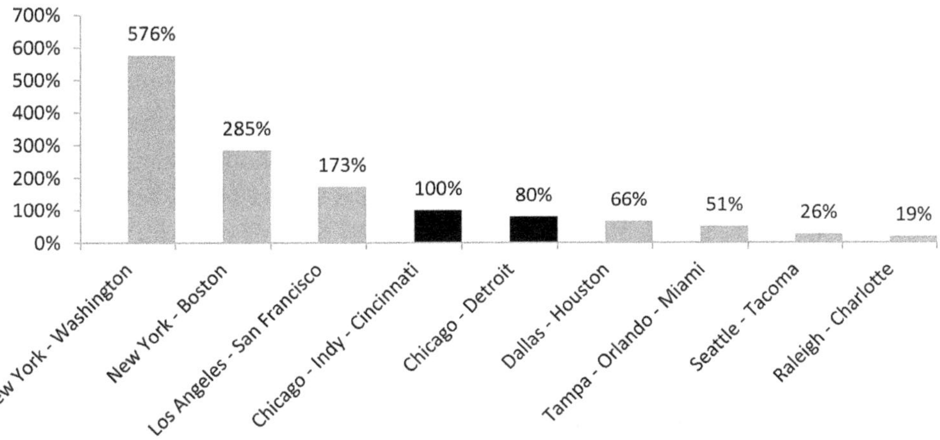

Dallas – Houston. Both are also appreciably larger than Miami – Tampa, even if the Orlando – Tampa segment (which was included due to its mileage) is added.

Passenger travel on the Midwestern routes is nevertheless less than a fifth of that of the New York – Washington corridor and about a third of Boston – New York. They have only about half as much travel as the California corridor. When the corridors are compared on the basis of *passenger miles*, the Midwestern Corridors are around 45% as long as the California corridors, which involve much longer travel distances.

When viewing corridors throughout the country, of course, the need to consideration additional variables beyond population and distance, such as affinity factors and political boundaries, becomes even more important. Even so, the exercise of evaluating corridors in this simplified way offers a compelling portrait of the size of the largest Midwestern routes.

FUTURE RESEARCH DIRECTIONS

The analysis presents above suggests that important research opportunities exist to foster a bettering understand the dynamics of intercity passenger service in the region.

1. Researchers should work to the fill the void in readily available data on the aggregate amount of travel between Midwestern cities. Filling this void would provide policy analysts with more credible baseline estimates for scenario analysis. Such research would be greatly facilitated by a resumption of the

American Travel Survey, which was last conducted by the federal government in 1995. This Survey, which was abandoned for budgetary reasons, could provide a benchmark that allows demand forecasters to better calibrate their models.

2. Additional research should explore factors excluded from the present study, such as: i) the affinity and tendency for traffic volume to be stronger on intra-state than interstate routes; ii) the importance of economic variables affecting travel volumes; and iii) the fact that the market served by rail service depends heavily on the length of the trip. This research could also consider the differing propensity for traffic on the corridors to divert to rail.
3. Research should be conducted on the potential synergy created by building a multi-city rail network that would allow for reliable and convenient connections between cities. There is a particular need to explore the extent to which an integrated Midwestern hub would provide for attractive linkages between many cities on opposite sides of the Chicago, much like an airline hub allow passengers to reach many destinations with a single change of planes. Creating such a hub would likely require major improvements to terminal facilities at Chicago Union Station (Schwieterman, 2014).

CONCLUSION

The methods used in this chapter, while only a start, provide a useful framework for understanding the relative market size of prospective Midwest high speed rail routes. The comparison of the number of *passengers* carried is most relevant when one considers such factors as the amount of congestion reduction in Chicago and the size of the constituency potentially served by rail improvements. *Passenger miles* are most relevant when considering greenhouse gas reductions, revenue generation, and economic development. Measurements of *traffic density* provide insights into the cost effectiveness of investments in rail-service.

Of course, estimates of the total size of the market may differ sharply from the size of the market for which high-speed rail service might be competitive. Nevertheless, despite its limitations, the gravity model is a simple and flexible tool than can help build understanding of the dynamics of corridors. Four corridors stand out as being appreciably larger than the rest. These are ranked as follows:

1. **Chicago – Indianapolis – Cincinnati**: This corridor is the only one of the seven that serves *three* metropolitan regions, each exceeding 1.5 million people. Under all three scenarios of the gravity model, it has the highest total traffic volume and is slightly ahead of Detroit on the basis of passenger miles.

Nevertheless, the large share of local traffic moving between Cincinnati and Indianapolis, in which travel distances are short, may render high-speed rail less attractive here than other modes. Another cautionary note: the MWRRS study estimates that traffic ranks much lower on this corridor.

2. **Chicago – Detroit**: This corridor ranks 2nd with respect to total passengers in all three gravity model scenarios and exceeds the Cincinnati route with respect to passenger miles in all but the low-friction scenario. The route boasts more than twice the endpoint-to-endpoint traffic as the others. If additional variables are considered, it might be found to be the largest market, although slower-than-average population growth in Michigan may limit future market growth. The route was determined to have the highest passenger volume in the MWRRS study.
3. **Chicago – Milwaukee – Twin Cities**: This corridor's large intermediate stops at Madison and Milwaukee add considerably to travel volumes, making it 3rd in passenger traffic and passenger miles in all three scenarios. The route attractiveness nonetheless falls when considering higher levels of friction due to the long travel distances, and is reported to be smaller than the Cleveland route in the MWRRS study. The inclusion of local Chicago – Milwaukee traffic would dramatically boost the estimated market size.
4. **Chicago –Toledo –Cleveland**: This corridor has the benefit of serving *four* cities, each with populations approaching or exceeding 500,000, something shared only by the Twin Cities routes. It ranks 4th in passenger traffic and passenger-miles but only 5th in travel between endpoints. The MWRRS analysis concluded its travel volumes exceeded the St. Louis and Twin Cities corridors by a considerable margin.
5. **Chicago – St. Louis**: The corridor consistently ranks 5th in both passengers and passenger miles in most scenarios. Its endpoint-to-endpoint traffic is nonetheless quite strong. The limited size of intermediate cities hampers its overall ranking.

Policymakers have placed considerable emphasis on the Detroit and St. Louis corridors for HSR development in recent years. Enhancing service to the Twin Cities route also been prominent on the public eye in part due to the controversial nature of rail proposals involving Wisconsin. The gravity model, however, suggests that the Chicago – Indianapolis/Cincinnati and Cleveland corridors also deserve heightened policymaker attention.

When viewed in totality, the Midwest stands out for having a wide variety of corridors, each with qualities that appear to make them promising for HSR development. Although it may be difficult to build a consensus about which route is best, there are good reasons to believe the above five corridors have the largest travel markets.

REFERENCES

Federal Railroad Administration. (2013). *Midwest Regional Investment Fact Sheet*. Washington, D.C.: U.S. Department of Transportation. Retrieved from http://www.fra.dot.gov/eLib/details/L04666

Hagler, Y., & Todorovich, P. (2013). *Where High-Speed Rail Works*, New York: American 2050. Retrieved from http://www.america2050.org/pdf/Where-HSR-Works-Best.pdf

Haynes, K. E., & Fotheringham, A. S. (1984). *Gravity and Spatial Interaction Models*. Beverly Hills: Sage Publications.

Midwest High Speed Rail Association. (2011). *The Economics of Impacts of High Speed Rail: Transforming the Midwest*. Midwest High Speed Rail Association: Chicago, IL. Retrieved from www.midwesthsr.org

Midwest Interstate Passenger Rail Commission. (2000). *Midwest Regional Rail System: Transportation System for the 21st Century*. Lombard, IL: Midwest Interstate Passenger Rail Commission.

Midwest Interstate Passenger Rail Commission. (2004). *Midwest Regional Rail Initiative Report*. Lombard, IL: Midwest Interstate Passenger Rail Commission.

Schwieterman, J. (2014). *Terminal Town: An Illustrated Guide to Chicago's Airport, Bus Depots, Train Station and Steamship Landings: 1939 - Present*. Lake Forest: Lake Forest College Press.

Schwieterman, J., & Scheidt, J. (2007). Survey of Current High-Speed Rail Planning Efforts in the United States. *Transportation Research Record: Journal of the Transportation Research Board, 1995*, 27–34. doi:10.3141/1995-04

TEMS, Inc. (2004). Midwest Regional Rail Initiative Project Notebook (Ch. 4).

KEY TERMS AND DEFINITIONS

City Pair: Departure (origin) and arrival (destination) cities comprising a transportation segment or travel itinerary.

Friction: The impedance or reluctance of persons to make trips of various distances. The general friction factor indicates that travelers are increasingly less likely to make trips as mileage increases.

Gravity Model: A model to estimate traffic volume based on the economic sizes (often using population or income measurements) and distance between two units. These models are derived loosely from principles of physics that underlie Newtonian laws of gravitational force.

Intermediate Cities: Cities location between the endpoints of transportation corridors. These are often called "intermediate stops."

Midwest Regional Rail System (MWRRS): A multistate rail system proposed by the Midwest Interstate Passenger Rail Commission, which has formal representation by various states in the Midwest region.

Passenger-Mile: A statistical unit denoting one *mile* traveled by one *passenger* and used by agencies of public transportation (as railroads, bus lines, or airlines) in measuring the volume of *passenger* traffic.

Travel Density: The average number of travelers per one mile or one kilometer of a transportation corridor, often expressed in passengers per mile.

Trip: The journey between one point and another.

ENDNOTES

1. TEMS Inc. (2004) Midwest Regional Rail Initiative Project Notebook, "Chapter 4 – Market Analysis: Midwest Regional Rail System, p. 15.
2. Traffic was increased annually by 1.5% to reflect the historical trend of transportation activity growing at roughly 2/3 the rate of GNP growth, which has averaged about 2% in the Midwest since 2000.
3. Much has been written about the relatively high airfares and problems facing Amtrak service to Cincinnati and Indianapolis. See especially Nate Silver, "Which Airports Have the Most UnFair Fares," dated April 6, 2001. Retrieved from http://fivethirtyeight.blogs.nytimes.com/2011/04/06/which-airports-have-the-most-unfair-fares/.
4. There is a possibility that the estimate of the amount of Cincinnati – Indianapolis traffic in the MWWRS study is too low. This study shows that it has only half the traffic as the Cleveland – Toledo route, which is a similar travel distance but has a combined population that is only about 60% as large as Cincinnati and Indianapolis.

APPENDIX

Table 3. Rankings of Midwestern corridor by scenario and variables measures

	Total Passenger Trips			Trips between Endpoints			Passenger Miles			Density		
	Low (1.3)	Medium (1.6)	High (1.9)	Low (1.3)	Medium (1.6)	High (1.9)	Low (1.3)	Medium (1.6)	High (1.9)	Low (1.3)	Medium (1.6)	High (1.9)
Chicago – Twin Cities	3rd	3rd	3rd	3rd	4th	4th	3rd	3rd	3rd	3rd	3rd	3rd
Chicago – Lansing	6th	6th	6th	6th	6th	6th	7th	7th	7th	6th	6th	6th
Chicago - Detroit	2nd	2nd	2nd	1st	1st	1st	1st	2nd	2nd	2nd	2nd	2nd
Chicago - Cleveland	4th	4th	4th	5th	5th	5th	4th	4th	4th	4th	4th	4th
Chicago - Cincinnati	1st	1st	1st	4th	4th	3rd	2nd	1st	1st	1st	1st	1st
Chicago - St. Louis	5th	5th	5th	2nd	2nd	2nd	5th	5th	5th	5th	5th	5th
Chicago - Omaha	7th	7th	7th	7th	7th	7th	6th	6th	6th	7th	7th	7th

Assumed rate of friction (traffic decline due to distance)

Chapter 4
Benefits Forgone:
The Cost of Not Building Higher Performing Passenger Rail

Eric C. Peterson
Transportation Policy Consultant, USA

ABSTRACT

By conservative, reliable, objective, and preliminary estimates, the United States could generate at least $660 million annually – $26.4 billion over four decades – by building and operating faster, more frequent, more dependable, and more highly integrated passenger rail service in four corridors: The Northeast; the Chicago Hub; California; and the Northwest. Numerous studies by a variety of national and international organizations and government agencies suggest that the actual cost of building and/or improving passenger rail service lines is significantly less than the cost per mile of other modal alternatives. In fact, in many corridors, passenger rail is the only feasible option for adding capacity, given the practical constraints facing aviation and highway expansion. Further, by integrating improved, higher speed intercity passenger rail service into the existing transportation system, major issues of congestion, mobility and economic inequality can be addressed.

INTRODUCTION

The U.S. Census Bureau estimates that the population of our nation will grow by more than 100 million over the next 40 years. A sizeable portion of America's present population is already searching for mobility alternatives, and demographers predict that desire will only continue to grow in the face of rising fuel costs, environmental

DOI: 10.4018/978-1-5225-0102-2.ch004

concerns, senior citizen mobility issues, and a steady growth in urban population centers brought on by Millennials and Empty Nesters (Davis, Dutzik, 2012; Baxandall, 2012; t4America, 2011).

All of these influences are putting pressure on America's transportation infrastructure, which, by most credible measures, is hampered by deferred maintenance and inadequate capacity. Since the early 1970s, transportation policy makers have recognized the need to maintain a competitive, multi-modal interurban transportation system. Unfortunately, the nation's investment and policy development has not kept pace with either demand or expectation, thereby constraining the transportation system and limiting consumer choice.

In the 1990s, beginning with the Clinton administration and continuing through the administration of George W. Bush, serious bipartisan efforts were undertaken to address the needs of the nation's transportation infrastructure – including its intercity passenger rail corridors. But midway into the first decade of the 21st Century a public policy paralysis took hold that virtually destroyed the bipartisan commitment to infrastructure. In the midst of this upheaval, both an outgoing administration and an incoming administration attempted to breathe new life into the policy discussion on intercity passenger rail in America. In 2008, President Bush signed into law the Passenger Rail Investment and Improvement Act (PRIIA), which authorized an aggressive set of initiatives that had the potential to redress the growing passenger rail mobility challenges facing the nation.

Under the dark clouds of one of the world's most severe economic crises, the new administration of President Obama pushed through Congress in 2009 a massive economic stimulus package containing more than $7 billion in rail investments and the outlines for a nationwide strategy to give "80% of Americans access to high-speed rail within the next 25 years" (Obama, 2011).

Between 2009 and 2011 Congress approved more than $13 billion in funding for intercity passenger rail improvements, and the Administration promised at least $50 billion more over the next four years. By the end of 2010, 39 states, the District of Columbia, and Amtrak submitted applications requesting more than $75 billion – well in excess of the available funding – for projects and corridor improvements in every region of the country.

But the Congressional elections of 2010 eliminated virtually all bipartisan supported and by the spring of 2011 Congress severely cut funding for the intercity passenger rail initiative, causing the Federal Railroad Administration (FRA) to disperse a little more than $10 billion in grant funding to 32 states and Amtrak. Additionally, Congress eliminated all future funding for the Administration's efforts to jump-start high-speed intercity passenger rail, and placed severe constraints on

Amtrak's ability to meet the growing demand for even its form of slower intercity and long-distance service.

In light of this history and the prospects for the future, policy makers are faced with the question: What is the cost of not building an improved intercity passenger rail system?

To answer this question, this chapter examines intercity passenger rail service from the perspective of a highly integrated transportation service that offers competitive door-to-door travel time comparable to other intercity transportation modes operating in particular corridors at distances of 100 to 600 miles.

This chapter addresses the cost to America's transportation system, and the nation's environment, economic condition, and energy supply if aggressive efforts are not made in the very near term to improve the U.S. intercity passenger rail system on par with the systems now operating in Europe and Asia.

Starting with an examination of the costs to the nation's commercial aviation system if improvements are not made to American intercity passenger rail, the chapter then addresses the cost to America's highway and roadway system. In each of these sections the capital, maintenance, environmental and social costs associated with expanding capacity to meet demand without improving intercity passenger rail will be identified.

The third section of this chapter addresses the cost to the nation's transit systems, including urban, suburban and rural transit if improvements are not made to the American intercity passenger rail system. The chapter concludes with an offering of perspectives that may lead to constructive policy discussions that could ultimately put American intercity passenger rail service on the right track.

Readers should recognize that this chapter is not an unabashed effort to advocate for high-speed or higher-performing intercity passenger rail. This chapter is simply an objective analysis of the costs and benefits of improving the nation's intercity passenger rail service in a manner that complements the rest of the nation's passenger transportation systems, offering opportunities to ease the congestion and stress travelers currently experience, congestion and stress that will continue to mount without other mobility options.

BACKGROUND

The underpinnings of this chapter are drawn from a September 1997 U.S. Department of Transportation (DOT) study entitled "High-Speed Ground Transportation for America" (U.S. Department of Transportation, 1997). To date, this study is one of the most, if not the only, comprehensive examinations of American intercity passenger rail development opportunities ever compiled. While exhaustive in its analysis, the 1997 study does contain several assumptions that time has shown to be inaccurate. For example, the 1997 study projected that fuel costs would not increase between 1997 and 2020. The study also did not contemplate, among other things, the growth in demand for alternative transportation modes among the Millennials – who did not exist at the time the study was written – and the Baby Boomers – who never considered that one day they might not want to, or could not drive. The 1997 study also did not consider the capacity and convenience changes that would come to the airline industry, the evolution of discount intercity bus service, or the capacity constraints now facing the American freight rail network – the backbone of the nation's commuter and intercity passenger programs.

Because many of the assumptions in the 1997 study have not held up over time, the conclusions and perspectives of that study may not be realistic today, but because they are the only ones available, they are, at least, a base line from which a rational discussion can go forward. And despite the fact that some of the 1997 study's assumptions have not survived the test of time, others have, and continue to shape the current and future direction of America's intercity passenger rail system.

Additionally, all dollar values reflected in the 1997 DOT analysis have been updated to 2012 values. Thus, when comparing dollar values in the 1997 DOT analysis to the dollar values shown in this chapter, the dollar values in this chapter have been adjusted to reflect that a dollar in 2000 at 2012 present value is $1.34.

This chapter only focuses on four of the 11 corridors originally discussed in the 1997 study. The Florida and Texas corridors included in the 1997 DOT analysis, as well as the Maglev technology corridors and technologies included in the 1997 DOT study, are not included in this chapter. Also, in the 1997 DOT analysis California was presented in two parts, the Chicago Hub was presented in three parts, and the corridors stretching from Montreal, Buffalo and Boston to Charlotte, NC were presented in three parts. In this chapter both sections of California are combined, as are all three portions of the Chicago Hub, and all three sections of the Northeast Corridor (U.S. Department of Transportation, 1997).

Additionally, this paper takes a different perspective on the issue of benefits by examining the cost of not enhancing the nation's passenger rail system, thereby continuing – as the United States has for the past half-century – to add and repair more lanes to highly congested roadways, and build more runways and add more flights to an already overcrowded and technologically outmoded aviation system.

Finally, this chapter takes a different perspective on the description and goals of the intercity passenger rail initiative previously characterized as "high-speed rail." As reflected in the Administration's 2013 budget for the U.S. Department of Transportation, this initiative will be described as HPPR – High-Performance Passenger Rail, with its goal to provide 80 percent of Americans access to a multi-modal transportation system that offers faster travel times, better reliability, more frequent service, and seamless connections to destination and other modes of transportation. This expanded and improved transportation network will help provide the additional capacity and travel options necessary to serve America's growing population (U.S. Department of Transportation 2013 Budget, 2012, p.22).

MAIN FOCUS OF CHAPTER

In essence, this chapter addresses the benefits forgone, i.e. the opportunity cost, of not building HPPR.

Opportunity cost is defined by the Paris-based Organization for Economic Cooperation and Development (OECD) as "…[a measure of] opportunities foregone at the time an asset or resource is used, as distinct from the costs incurred at some time in the past to acquire the asset, or the payments which could be realized by an alternative use of a resource, e.g. the use of labor in a voluntary capacity being valued at the wages which could have been earned in a paid job" (OECD, 2001, http://esa.un.org/unsd/sna1993/introduction.asp).

Included in this discussion are the capital, maintenance, environmental and social benefits that may not be realized, and an extrapolation of the costs the nation may incur if nothing, or very little, is done to enhance America's intercity passenger rail system, thereby not allowing HPPR to become a viable option for future travelers.

Admittedly, given the history of passenger rail in the United States, many find it somewhat unrealistic to consider that passenger rail at any speed or configuration has a sustainable future in America. Yet, continued intercity passenger service ridership growth and demographic changes offer the possibility of a more optimistic perspective. This chapter offers an objective, balanced view of what the costs and benefits of improving America's intercity passenger rail system may be, especially the benefits forgone if nothing is done.

AVIATION BENEFITS FOREGONE FROM NOT BUILDING HPPR

America's commercial aviation system faces significant challenges. The Airport/Airway Trust Fund that is supposed to support the development, maintenance and operation of the nation's aviation system is so strapped that it must rely on the U.S. Treasury for more than 35 percent of its revenue, and the system is facing an $80.1 billion (annualized at $16 billion) capital development backlog between now and

2015 according to Airports Council International – North America (Airports Council International/North America, 2011).

These capital development projects include:

- Expanding an airport's capacity beyond its current design to meet growth in demand for aviation services
- Bringing an airport up to FAA-mandated design standards to achieve full productivity of aircraft using the airport
- Reconstructing aging airport infrastructure
- Upgrading infrastructure to accommodate the introduction of different aircraft types
- Addressing safety, security and environmental concerns (Airports Council International/North America, 2011).

Included in this backlog are 40 percent of all airside projects, 45 percent of all terminal projects, and 16 percent of landside projects that cannot proceed due to inadequate funding. Over all this represents approximately 30 percent of all projects at large hub airports, nearly 40 percent of all projects at medium hub airports, and about 25 percent of all projects at small hub airports (Airports Council International/North America, 2011).

Among recent airport expansion projects that have been completed are a new fifth 6,000 ft runway at Hartsfield Airport in Atlanta, Georgia that opened in 2006 at a cost of approximately $1.28 billion (Atlanta Constitution, March 31, 2006), and in 2008 a new fourth runway opened at Washington's Dulles International Airport at a cost of $49.7 million (Metropolitan Washington Airports Authority, www.washingtonairports.com/dulles/773).

The Florida Department of Transportation estimated that the cost of constructing airport runways and complementary facilities can cost well beyond $2,200 per linear foot (Florida Department of Transportation 2011,).

The planned expansion and renovation of Chicago's O'Hare Airport was estimated in 2010 to be $6.6 billion and adding 16 new gates at the Minneapolis-St. Paul Airport was expected to cost approximately $400 million (Wisconsin Public Interest Research Foundation, 2010).

These are the fortunate few. There are many airports where expansion is not logistically or politically possible, but population growth will continue to strain mobility capabilities (Airports Council International/North America, 2011).

Added to the backlog and cost of airport construction, the Federal Aviation Administration (FAA) has been struggling for 20 years to put in place a $40 billion navigation system upgrade that was originally targeted to come on line in 2016. Now it appears unlikely that the new system, known as NextGen, will be operational before 2025. Holding up NextGen's progress are both institutional and funding challenges, even though the new system is intended to reduce costs and increase the capacity of the nation's airways.

Ironically, if and when NextGen is implemented, it will have the potential of compounding the airside congestion that most of the major airport hubs and several of the regional airport hubs are experiencing (ENO Center for Transportation, 2012).

Ball, Barnhart, Dresner, et al (2010) estimated that due to current conditions on the ground, U.S. transportation delays in 2007 reached $32.9 billion. Approximately 20 to 35 percent of all domestic airline flights are delayed by more than 15 minutes (the average is 59 minutes), according to the U.S. DOT's Bureau of Transportation Statistics (BTS). These delays cost the airlines over 122 million hours of delays in 2010, at a financial loss of more than $3.5 billion annually (Bureau of Transportation Statistics, Research and Innovative Technology Administration, U.S. Department of Transportation, 2011).

Together these losses and unaddressed costs will total somewhere between $137.5 billion ($27.5 billion annually) and $284.8 billion ($56.9 billion annually) over the next five years (Airports Council International/North America, 2011; Bureau of Transportation Statistics, Research and Innovative Technology Administration, U.S. Department of Transportation, 2011; ENO Center for Transportation 2012; Ball, Barnhart, Dresner, et al, 2010).

The combination of population growth and no viable alternative mode to help offset the anticipated growth in air travel forecast by the FAA will cause congestion and its associated costs to become even more acute for the nation's airlines and airports. While not yet back to the record passenger levels of 2008, the number of enplanements on mainline and regional air carriers grew to 739.3 million in the United States in 2013 and FAA expects that number to grow to as much a 1.15 billion by 2034 (FAA, 2014).

Add to that a projected near tripling of total air cargo (freight/express and mail) from 34.8B revenue ton miles (RTM) in 2013 to 81.0B RTMs in 2034 and it is apparent that the nation's aviation network as well as the airside capacity of the nation's airports will be in need of the kind of mobility capacity HPPR could offer (FAA, 2014).

Throughout the nation are corridors that stretch between 100 and 600 miles in which airline travel is seriously congested, and for which HPPR could be highly competitive alternative from both time and pricing perspectives (Machalaba, 2007, p. B1).

The Passenger Rail Working Group of the National Surface Transportation Policy and Revenue Study Commission, a congressionally chartered industry advisory body, identified eight major metropolitan corridors in the United States where 14 airports will require non-aviation mobility support (i.e., increased and/or new passenger rail service) even if planned aviation improvements are made by 2025 (Passenger Rail Working Group of the National Surface Transportation Policy and Revenue Study Commission, 2007, pp. 11 – 12).

In two of these corridors, Washington/New York and New York/Boston, Amtrak already carries as many passengers daily as the scheduled airlines. In addition, with more frequent, reliable, and marginally faster service, intercity passenger rail could be an attractive alternative to the hassle, cost, and travel time of commercial aviation in the Washington/Richmond, Virginia; Los Angeles, California/San Diego, California; Portland, Oregon/Seattle, Washington; and Chicago, Illinois/Milwaukee, Wisconsin corridors ((GAO, March 2009; Passenger Rail Working Group of the National Surface Transportation Policy and Revenue Study Commission, 2007, p. 12).

The experiences of the Europeans and Japanese suggest that HPPR could bring about significant mode-of-travel shifts in certain U.S. corridors. For example, in France air travel in the TGV Sub-Est corridor went from 31 percent of the market to just seven percent after the introduction of the TGV. In the Madrid/Seville corridor air travel shifted from 40 percent of the travel market to 13 percent following the introduction of AVE (Nash, 2007).

Amos, Bullock, and Sondhi (2010) observed in their July 2010 World Bank report that airline service has been suspended altogether in many of the shorter European and Japanese corridors, thus freeing up airlines to redeploy their fleets to longer, more lucrative city-pairs while reducing congestion at other regional hubs (Amos, Bullock, and Sondhi, 2010).

The 1997 U.S. DOT study projected that, depending on the corridor, the speed, and the travel time door-to-door, HPPR could divert between 16 percent and 30.9 percent of airline passengers, producing total airport congestion delay savings of approximately $19 billion annually in the four corridors this chapter examines. Those are savings that would be forgone without HPPR (U.S. Department of Transportation, 1997).

Additionally, the 1997 DOT study demonstrated that HPPR can deliver significant mobility benefits for travelers wishing a transportation alternative to the airlines, and an overall transportation benefit that rewards both the airlines and the passengers who may continue using the airlines as their preferred mode of travel. These are benefits that would be foregone if HPPR is not built.

In the California corridor from Los Angeles to San Francisco, for example, DOT estimated that for new high-speed rail operating on almost completely new right-of-way with maximum practical operating speeds of 200 mph, the mode shift from air to rail would be approximately 24.4 percent, or approximately 39 percent of the HPPR ridership. The total airport congestion delay savings in the four corridors addressed in this chapter would be almost $16.27 billion ($21.8 billion in 2012 dollars) annually over the next ten years. Such savings would result in a net reduction of air travel delay costs ranging from approximately one third to nearly four times airports' annual delay costs depending on which estimate of costs is used (U.S. Department of Transportation, 1997).

More recently Haas (2014) reported that a secondary analyses of previous studies and in newer studies that use primarily original data, that the presence of HSR in Europe and Japan has resulted in significant-to-dramatic mode shifts. The studies reviewed comprise a large variety of HSR systems, time periods, data sources, and means of analysis (Haas, 2014, p. 1).

Energy and Environmental Costs

Beyond the congestion cost savings HPPR can generate for commercial aviation, the estimated fuel savings of HPPR on a per-passenger-mile basis compared to flying are quite significant. According to American Association of State Highway and Transportation Officials (AASHTO, 2014, website) even today's trains are 17 percent more fuel efficient than airplanes on a per-passenger-mile basis. This figure is supported by a 2009 paper presented to the 18th International Transport Research Symposium of the OECD that demonstrated that with a 70 percent load factor, energy consumption per seat kilometer for aircraft is 2.57 megajoules (MJ) per p/km (a megajoule is equivalent to one-third of a kilowatt hour of energy) compared to 0.5 MJ per p/km for conventional trains and 0.76 MJ per p/km for high-speed trains (Nash, 2009, p 8).

If one assumes that 45 percent of travel in a given corridor is air travel diverted to HPPR, as has happened in the Paris/Madrid corridor, then the energy savings can be rather large. Drawing on data in the 1997 DOT study, the emissions savings in the Northeast Corridor alone could be over $404 million a year. Given the energy savings of HPPR compared to aviation, it follows that the environmental impact per passenger and passenger mile is almost as dramatic.

Depending on the power source for HPPR, the European experience suggests that HPPR could produce as much as 6.5 times less particulate matter per passenger and 31.9 times less nitrogen oxides per passenger than aircraft over the same distance (Dutzik T. & Steva E., 2010 pp. 14 -18).

Kageson (2009) estimated that even with the anticipated technological improvements that will come to the various transportation modes over the next 10 to 15 years, the amount of emissions of greenhouse gases per seat kilometer will still advantage HPPR over aircraft. A train running at speeds up to 150 km/h will emit 9.5 grams of carbon dioxides per seat kilometer, and a train running at speeds up to 280 km/h will emit 15.4 grams of carbon dioxides per seat kilometer compared to an aircraft emitting 93.8 grams of carbon dioxides per seat kilometer. Assuming load factors of 80 percent for regional aircraft, 75 percent for high-speed trains, and 65 percent for conventional intercity trains, Kageson estimated the grams of carbon dioxide emissions per kilometer to be 14.6 for conventional trains, 20.6 for high-speed trains, and 117.2 for regional aircraft (Kageson, 2009, pp 5, 11 – 12, 17 - 18).

Without HPPR, it is clear that the impact of aircraft emissions on the environment will continue to be significant despite the fuel efficiency and emissions improvements airplane manufacturers will make over the next 10 to 15 years (Jehanno A., James C., & Palmer D., 2011, pp 16 -17),

Social Costs

As the nation's only intercity passenger rail enterprise, Amtrak provides service to over 500 communities across the United States. A large percent if its ridership is individuals who are unable to fly or drive. These are senior citizens (who account for 38 percent of adult passengers) and disabled persons. In fact, Amtrak notes that 42 percent of passengers with disabilities who traveled on Amtrak in fiscal year 2010 rode long-distance trains. Additionally, Amtrak is often the only transportation mode still operating during severe winter weather conditions that ground planes and close highways, and plays an important role in emergency situations like after the terrorist attacks of September 11, 2001 when the U.S. government grounded all air travel (Amtrak, 2014, http://www.amtrak.com/ccurl/374/876/Long%20Distance%20Trains.pdf).

In comparison to air travel, the Government Accountability Office (2009) notes that at distances of 100 to 500 miles, higher speed passenger rail can be more convenient, especially if travel is from one city center to another because of reductions in access and waiting times.

In the next section of this chapter HPPR will be examined for its potential to improve mobility by serving as an alternative travel option. This is particularly important at a time when the airlines face capacity constraints and rising fuels costs, and are either reducing or eliminating air service to many communities altogether. The top 50 airports in the nation now represent more than 80 percent of all airline departures. At the same time more than half of those airports had fewer flights in 2010 than they did in 2009 (Mouijawad 2011).

Finally, if one accepts that the policy objective for federal funding of the aviation sector is to strengthen the Airports/Airway Trust Fund in order to reduce and/or eliminate its reliance on general fund revenue, thereby securing adequate reliable revenue to fully implement the NextGen navigation system and to reduce the airports/airways capital development backlog, then one should also accept that neither the commercial aviation sector nor the traveling public would be well served if HPPR is not constructed as part of a highly integrated national transportation system.

HIGHWAYS AND ROADWAYS CAPITAL AND MAINTENANCE COSTS

Similar to the conditions facing the nation's aviation sector, the highway and roadway sector faces huge investment and congestion challenges that, despite their widely publicized impact on the productivity and competitiveness of the nation, have not been adequately addressed by policy makers.

This section examines the capital and maintenance, energy and environmental, and social costs associated with highway and road construction, operation, and maintenance if HPPR is not built as part of a highly integrated transportation system.

Because of the deteriorating condition of the nation's roadways and the growing demand for their use, the American Society of Civil Engineers (ASCE) annually releases a report card on the condition of the nation's transportation infrastructure. In recent years ASCE has given our nation's transportation infrastructure a "D," on par with the developing nations of the world. The latest report projected that the condition of the nation's surface transportation system would "cost the economy more than 870,000 jobs and suppress the growth of the country's gross domestic product (GDP) by $3.6 trillion by 2020" (American Society of Civil Engineers, 2014, website).

The Federal Highway Administration's (FHWA) (2013) "Status of the Nation's Highways, Bridges, and Transit Conditions and Performance," reported that in order to just maintain the nation's surface transportation network at 2010 levels through 2030, the U.S. would need to invest between $65.3 billion to $86.3 billion per year. If major improvements were desired, however, FHWA estimates the annual investment would be $123.7 billion to $145.9 billion per year.

On top of that, FHWA estimates that in order to bringing existing transit assets up to a state of good repair the nation would need to invest, at a minimum, about $18.5 billion annually through the year 2030 (Federal Highway Administration 2013).

Because of the fiscal constraints of the Federal Highway Trust Fund and the difficulty faced by Congress and the Administration in passing regular surface transportation authorizations and appropriations, in 2005, under the Safe, Accountable, Flexible, Efficient Transportation Equity Act: A Legacy for Users (SAFETEA-LU), two commissions were authorized to study the financing and structure of the federal government's surface transportation program. One of the authorized commissions, The National Surface Transportation Infrastructure Financing Commission (2008), projected in its report, "Paying Our Way: A New Framework for Transportation," that absent a change in funding policy, an annual gap between the funding needs of our nation's highways and roads and the available revenue (including the federal motor fuels tax) would be in the range of $134 billion to $194 billion a year for the period of 2008 to 2035 (The National Surface Transportation Infrastructure Financing Commission, 2008, p.2).

Investment levels included in the subsequently enacted Moving Ahead for Progress in the 21st Century (MAP-21) did not arrest the deferred maintenance and capacity needs.

Additional signs of the ominous condition of the nation's roadways are reflected in the annual "Urban Mobility Report" of the Texas Transportation Institute (TTI) at Texas A&M University that annually analyzes traffic congestion in urban areas across the United States and identifies both the time consumed and the volume of fuel burned in the course of the nation's daily commute.

The latest TTI report painted a bleak picture of seemingly endless "rush hours," massive congestion delays, lost worker productivity, and billions of gallons of wasted fuel as a result of drivers idling in gridlock. Highway and roadway congestion cost the nation more than $101 billion in wasted fuel and lost productivity in 2010. Intercity passenger travel by car was both a contributor to and a victim of this congestion (Texas Transportation Institute, Texas A&M University, 2011).

Looking forward through 2015, the 2011 TTI study predicted that commuters will experience an additional three hours of annual delay (up to 37 hours a year) with a total annual national cost in lost productivity of more than $133 billion – more than $900 per commuter, and worst of all, the annual loss of more than 2.5 billion gallons, adding to concerns over national security, economic vitality, air quality and needless greenhouse emissions.

One strategy for addressing highway/roadway congestion might be to build more highways and roadways. Clearly some will need to be built and/or reconditioned. The American Road & Transportation Builders Association (ARTBA) (2014, www.artba.org/About/transportationfaqs/#20) suggests the following costs for roadway construction and reconditioning:

- Construct a new 2-lane undivided road – about $2-$3 million per mile in rural areas, about $3-5 million in urban areas.
- Construct a new 4-lane highway — $4-$6 million per mile in rural and suburban areas, $8-$10 million per mile in urban areas.
- Construct a new 6-lane Interstate highway – about $7 million per mile in rural areas, $11 million or more per mile in urban areas.
- Mill and resurface a 4-lane road – about $1.25 million per mile.
- Expand an Interstate Highway from 4 lanes to 6 lanes – about $4 million per mile.

That compares with an average rail construction cost per km of € 8 million (euros); about $16.8 million per mile, plus an additional 10 percent for land and planning costs, but does not include the cost of building stations. Bringing the total construction cost to approximately € 9.9 million per km; about $21.4 million per mile (De Rus, 2008 p. 10).

But these rail construction costs are off-set by the fact that rail right-of-way consumes about one-third of the land roadways require. And because its focus is city center to city center, HPPR tends to support smart growth principles that encourage denser infill development and less sprawl as well as other environmental benefits that will be discussed later in this chapter (Neff, 2011, pp. 28 -29).

Viewing HPPR as a means of reducing congestion and easing the capacity crisis on the nation's highways, the U.S. Department of Transportation (1997) estimated that in its first years of operation HPPR might divert as little as 1.1 percent to as much as 6.3 percent of intercity automobile travel in the four mega-corridors examined in this study.

Over time, as ridership in each of the HPPR corridors grows, the percent of intercity automobile traffic diverted to HPPR will also likely grow. Indeed, as Haas (2014, pp. 7 - 8) notes, in Europe the mode shift from car and bus to HPPR affirms this expectation.

As well, Nash (nd) observed that car and bus travel in the TGV Sud-Est corridor shifted from 29 percent before the introduction of the TGV to 21 percent afterward. In the Madrid-Seville corridor the shift was from 44 percent before the introduction of the AVE to 36 percent afterward. Japan, Korea and Taiwan experienced even greater shifts suggesting that as much as 25 percent to 45 percent of intercity automobile travel could be diverted to HPPR.

Such a diversion of intercity automobile travel could have dramatic impact on the ability of states and localities to maintain their roadways, and could significantly alleviate roadway congestion. The U.S. DOT suggests the savings just from reduced highway delays could be in the range of $489 million annually to $2.9 billion annually depending on the corridor. Those are savings that would be forgone without HPPR (U.S. Department of Transportation, 1997).

For example, in the Los Angeles to San Francisco portion of the California corridor, U.S. DOT (1997) estimated that 6.3 percent of drivers would shift to HPPR and in the San Diego to Los Angles portion of the corridor 1.1 percent of drivers would shift from autos to HPPR. In the four mega-corridors examined by the U.S. DOT the projected mode shift from auto to HPPR over all was on average about 3.1 percent, representing approximately 23 percent of the HPPR ridership (U.S. Department of Transportation, 1997).

By comparison, the Passenger Rail Working Group, National Surface Transportation Policy and Revenue Study Commission (2007, p. 37) projected 8.2 billion to 46.7 billion miles of annual vehicle travel savings with the development of HPPR, at a cost savings of $.7 billion to $6.6 billion annually.

These projections were based on an assumed 45 percent average load factor with a majority of riders diverted primarily from highway travel, producing travel-time savings valued at $11.20 per hour, based on US DOT's Transit Economic Requirements Model (TERM) (The Passenger Rail Working Group, National Surface Transportation Policy and Revenue Study Commission, 2007, p. 37).

These two studies, commissioned ten years apart, demonstrate that HPPR can deliver significant mobility benefits for travelers wishing a transportation alternative to driving and overall transportation benefits that reward both the HPPR traveler and the passengers who may continue using the highway as their preferred travel mode. These are benefits that would be foregone if HPPR is not built.

Environment and Energy Costs

The Passenger Rail Working Group (2007) report, also estimated the fuel savings that would be derived by the modal shift from car to HPPR would range from $.04 billion annually to $2.2 billion annually. These projections were based on a 45 percent average load factor for filled passenger seats during operations and assumed that train passengers would be primarily diverted from highways, based on US DOT's Transit Economic Requirements Model (TERM) (The Passenger Rail Working Group of the National Surface Transportation Policy and Revenue Study Commission, 2007, p.37).

On Europe's London to Paris corridor, it is estimated that passengers use one-third as much energy traveling by train as by car, and in the Madrid to Barcelona corridor, the fuel consumption is 28 percent of car travel. In Japan, train travel consumes, per passenger mile, one-sixth that of cars, and even more astounding is that the latest high-speed trains are 32 percent more energy efficient than the original high-speed trains while operating at speeds 43 miles per hour faster than the original Shinkansen trains (Dutzik & Steva, 2010, pp. 12 - 14).

From an energy perspective, the cost of not building HPPR reflects the fact that there is no other viable transportation alternative that can reduce the demand for fossil fuel so dramatically. Depending on the power source for HPPR, i.e., electrification or internal combustion, the amount of fuel savings will vary, but just based on current comparisons of the nation's largely internal combustion powered intercity and commuter passenger rail system, it is already recognized that per passenger mile, train travel is 21 percent more fuel efficient than auto travel. As the amount of train travel powered by electricity increases, the level of fuel efficiency will continue to increase.

With transportation currently consuming about 60 percent of the nation's energy demand, as long as that demand stays at current or higher levels, our nation will continue to face unacceptably high levels of economic and security risk.

The U.S. DOT study (1997) did not attempt to calculate fuel savings per se, but did attempt to calculate emissions savings from the diversion of auto travel to HPPR. In its study, DOT anticipated emissions savings based on the projected percent of highway traffic diversion in each corridor. For example, in the Northeast Corridor total emissions savings are estimated at $667 million annually with approximately 2.6 percent of highway traffic diverting to HPPR (U.S. Department of Transportation, 1997).

Santini D., Wang M. & Warner S., (1994, pp. 42 - 57), valued nitrous oxide emissions in Los Angeles at $26,400 per ton, carbon monoxide at $9,300 per ton, reactive organic gases at $18,900 per ton, and particulate matter at $4,700 per ton. Based on the experience of train service between Frankfurt and Basil, and Paris and Marseille, particulate matter per passenger was approximately 18.1 times less than automobiles, and nitrogen oxides were 46.2 times less per passenger (Dutzik T. & Steva E., 2010, pp. 16 - 19).

Thus, depending to the power source for HPPR, the environmental impact of market share shifting from auto to HPPR could be quite significant. If HPPR is not built, these are environmental savings foregone.

Social Costs

In the face of these considerations, the 2010 Census and the Federal Highway Administration's 2009 National Household Travel Survey point to changes in commuting and transportation habits among both the oldest and the youngest cohorts of the nation's population, particularly those in the 16 to 34 and the 50 and over age groups, that may provide a glimmer of hope for even more significant shifts from highway travel to HPPR. Both age groups are expressing strong desires for a more balanced transportation system with attractive options for alternative transportation modes like mass transit, including HPPR, to meet their commuting and intercity traveling needs. Additionally, the 2009 National Household Travel Survey reported that between 2001 and 2009, the number and percent of households with no vehicle available grew by nearly one million households, from 8.1 percent of all households to 8.7 percent (Federal Highway Administration, U.S. Department of Transportation, 2009 pp. 33 - 41).

The annual vehicle miles traveled per person peaked at around 10,000 in 2004, while total number of vehicles on the road plateaued in 2006. Additionally, as a share of the population eligible to have driver's licenses, fewer young people are obtaining them. Americans who do not drive increased by 1.1 million over the past 10 years with about 12 percent of adults older than 65 and nearly 30 percent of those over 75 no longer driving themselves. These are trends similar to travel choices being made in Europe, Japan and other countries that today operate a mix of public transportation options including HPPR (Sivak & Schoet 2011, pp. 14 - 15; Baxandall, Davis & Dutzik, 2012).

Based on the analysis of the Passenger Rail Working Group (2007) and projected in 2012 dollars, the nation would forgo approximately $15.16 billion in congestion delay savings, annual time savings, annual fuel savings and emission savings on U.S. highways and roadways if HPPR is not built.

So while the mode shift from auto to HPPR may not seem significant as a percentage of either travelers shifting to HPPR or percentage of HPPR riders, the overall savings in terms of congestion, time, fuel and emissions are indeed quite significant. Thus, when these savings are combined with both the size of the maintenance and construction backlog and the current and future demographic changes, the cost of not building HPPR is quite large, especially with regard to the future mobility of our nation, its citizens, its commerce, and its security.

PUBLIC TRANSPORTATION AND ECONOMIC DEVELOPMENT BENEFITS FROM HPPR

Public transportation and HPPR are highly complementary, and if integrated can provide a seamless means to travel from an origin to a destination conveniently and economically. Together, HPPR and transit are a welcome combination that improves mobility, and enables the traveler to maximize the use of his/her time while having minimal impact on the environment and limiting the use of environmentally damaging fossil fuels.

As Paris, London and Tokyo, to name a few, have experienced, HPPR is simply the longer distance, higher speed extension of a transit system that may include community circulators, cross-town bus service, bus rapid transit, trolley, light rail and/or heavy rail service, and/or commuter rail service. The linkage of transit to HPPR, just as the linkage of transit to aviation and roadways, adds value to both modes, enhances the mobility of travelers, and offers the opportunity for greater economic development than any single mode by itself could generate. The cost of not building HPPR as part of a highly integrated transportation system is the loss of these benefits (Dutzik & Steva, 2010 pp 2 – 4, 21 – 27).

Henry and Marsh (2009) observed that:

Promoting intermodal coordination among various public transport services has been a nominal goal of federal policy at least since the passage of the groundbreaking Intermodal Surface Transportation Efficiency Act (ISTEA) of 1991.

They noted that the National Commission on Intermodal Transportation (NCIT) – a body established under ISTEA – identified intermodal connectivity in public transport as a challenging but critical need and that, "Passenger intermodalism has shown some signs of progress since passage of ISTEA" (Henry & Marsh, 2009).

Chief among Henry and Marsh's findings were that integrated transportation systems generate greater efficiencies through cost sharing, smoother intermodal interfaces, greater local, regional and intercity connectivity, greater mobility choices, and stronger incentives for successful transit oriented development (TOD) (Henry & Marsh, 2009).

In its 2007 report to the National Surface Transportation Policy and Revenue Study Commission, the Passenger Rail Working Group noted that the anticipated demographic changes of the American population and the resurgence of urban centers suggest that public transportation and intercity passenger rail will play an important role for the growing segments of population that cannot or choose not to drive or fly (Passenger Rail Working Group of the National Surface Transportation Policy and Revenue Study Commission, 2007 p. 27).

According to the American Public Transportation Association (APTA), "Americans took 10.7 billion trips on public transportation in 2013, the highest annual ridership since 1957 (American Public Transportation Association, 2014).

Interestingly, while virtually all areas of the nation witnessed this growth, many of the leading growth areas were also areas that are promoting improvements to their intercity passenger rail service as well. This is a clear indicator that the demographic trend is moving back toward more highly integrated public transportation.

At the same time, the age cohort younger than 35, according to the latest U.S. National Travel Household Survey, prefer a transportation system that provides good options and choices for their various trips. They are more inclined, where possible, to rely on public transportation, intercity passenger rail, and the airways to serve their transportation needs (U.S. Department of Transportation, Federal Highway Administration, 2009 pp. 54 - 56).

Indeed, the 2009 National Household Travel Survey shows that since 2001 the share of trips in private vehicles has dropped while the share of trips on public transportation has grown for the age groups of 16 – 49 and 65+. For those who are not conveniently situated near public transportation, including intercity passenger rail, mobility has been significantly impaired. Still, according to Transportation for America, in 2009, seniors accounted for 9.6 percent of the more than 10.3 billion trips taken on public transportation in the United States. Over all, seniors made 328 million additional trips by transit in 2009 compared to 2001 (Transportation for America, 2011 pp. 11 - 14).

Additionally, Amtrak continues to enjoy steady growth in ridership. In FY 2013, a record 31.6 million passengers rode Amtrak on more than 300 daily trains at speeds up to 150 mph (241 kph) (Amtrak, 2014).

To take full advantage of the integration of HPPR with public transportation, cities like London have pursued regeneration strategies that promote agglomeration – the creation of catchment regions that extend the reach of urban transit systems that link economic activities to labor markets (Greengauge 21, 2006; Systra 2009).

Agglomerative strategies also produce environmental benefits. Even at existing levels of modal integration in 1997, the Federal Transit Administration (FTA) found that levels of transit use annually avoid about 125 million pounds of hydrocarbons and over 150 million pounds of nitrous oxides that would otherwise be emitted by automobiles (U.S. Department of Transportation, 1997).

Transportation accounts for nearly one-third of U.S. carbon dioxide emissions. According to the American Public Transportation Association (APTA), public transportation in the United States saves 4.2 billion gallons of gasoline and about 37 million metric tons of carbon dioxide annually. Note that this is based on existing levels of transit use. There will be additional efficiencies as ridership trends continue.

Building HPPR that is highly integrated with other transportation modes, most especially public transit, offers the United States the opportunity to launch an economic revolution that can offset all the disadvantages and challenges that beset the nation due to crippling congestion on its highways, bewildering delays and even loss of air service at its airports, and the demographic dynamics of an aging and increasingly urban population.

The objective is not to put HPPR in competition with other modes of transportation, but to establish HPPR as a complementary mode, offering travelers another option for mobility.

As documented in other sections of this chapter, improvements in intercity passenger rail, in terms of speed, frequency and reliability, have had dramatic impact on mode choices and demand. Additionally, many countries with improved intercity passenger rail service have experienced growth in the numbers of induced travelers – people who may not have traveled before by any mode, but have found HPPR a welcome travel option (Australian Government, Department of Infrastructure, Transport, Regional Development and Local Government, 2010).

de Rus (2008) offers a "simple cost-benefit model for evaluation of high-speed rail" based on the notion of doing something, i.e. introducing transportation alternatives. Conceivably, this same model could be used to evaluate HPPR, because ultimately the objective is to evolve ridership to a level that justifies high-speed rail investment (de Rus, 2008 pp. 14 – 16).

Because there is little data available, it is difficult to identify with dependable specificity the actual economic benefit to the nation and to the various regions in which HPPR may be pursued. In the alternative, understanding what these benefits might be, and thus what the cost of not building HPPR might be, is to look at the potential impact of eliminating existing intercity passenger rail service (Passenger Rail Working Group of the National Surface Transportation Policy and Revenue Study Commission, 2007 pp. 39 - 40).

In their paper, "Economic Impacts of Intercity Passenger Rail Service: Evidence from Passenger Surveys," Morgan and Sperry (2011) apply a direct measurement technique to both project future benefits if a particular corridor is enhanced by HPPR, as well as identify the impact if current service in the corridor was discontinued.

They observe that based on passenger surveys administered to passengers on the Heartland Flyer – a 206-mile Amtrak passenger train that operates daily round trip service between Oklahoma City and Fort Worth – 70 percent of the riders would divert their trips to other modes of transportation while fully 30 percent of the ridership simply would not make the trip at all.

In addition to the likely 39,000 new annual auto trips that would be put on the roadways in this corridor, the economic impact of that loss of travel as measured by passenger spending and the sales tax revenue generated by that spending in each of the communities served by the Heartland Flyer is fully one-third of the total spending, and one-fourth of the sales tax revenue generated by the ridership of the Heartland Flyer. For the corridor and the communities within it that loss is an estimated $1 million in annual reduced passenger spending and nearly $381,000 in annual lost sales tax revenue. Project this impact across the four mega-corridors discussed in this chapter and the cost of potentially lost service as well as the future economic impact of not building HPPR could be very significant (Sperry & Morgan, 2011 pp. 27 - 29).

A similar study done for Amtrak of the economic impact of its Empire Builder service in Montana estimated nearly $14 million in annual economic benefits. These are benefits that would be forgone without intercity passenger rail service (Amtrak website, 2013).

FUTURE RESEARCH DIRECTION

As noted from the outset of this chapter, projected costs of not building HPPR are based on assumptions and findings of the U.S. Department of Transportation's 1997 study, "High-Speed Ground Transportation," with adjustments made to reflect 2012 dollar values, and references to more recent research that either validates or disputes the 1997 study's findings. The reason for using the 1997 study is that it is the most comprehensive analysis available to date.

On this basis, it is clear that more current research is needed that relies on assumptions and findings that more closely reflect economic, demographic, environmental and other factors, as they exist almost 20 years after the initial Department of Transportation study.

Such research is needed for a variety of reasons, the most important of which may be to provide an objective body of credible data with which policy makers can factually determine the societal value of a publicly supported, reliable intercity passenger rail system that offers competitive travel times and significant economic and environmental advantages to other transportation alternatives.

A research project of this nature could be executed through a number of channels including through the Federal Railroad Administration's Office of Planning and Development, through the U.S. Department of Transportation's University Transportation Center research grant program, through the National Cooperative Railroad Research Program, or through the financial support of other private sector organizations and interests that give voice to concerns about the level of development, maintenance, operation, and competitiveness of America's transportation system.

CONCLUSION

The 1997 DOT study provides the best overview of the potential net benefits that could flow from the construction and operation of HPPR in the United States. Tables 1 and 2 reflect the total benefits, minus the total average cost of building HPPR in the four mega-corridors studied in this chapter. The net benefit in each of the four corridors reflects the total projected 40-year cost of not building HPPR. This table, however, does not reflect many of the social and agglomerative costs discussed in

Table 1. The benefits of building HPPR (millions of USD—2012 present value)

	Highway Delay Savings	Road Costs to Achieve Savings	Emissions Savings	Airport Delay Savings	FAA Spending Savings	HSR User Benefit	Transportation Benefits
California	$3,237	$12,950	$966	$9,908	$4,652	$20,900	$52,613
Chicago Hub	$927	$3,709	$154	$2,385	$1,120	$5,620	$13,915
Northeast Corridor	$3,857	$15,426	$667	$3,815	$1,791	$55,949	$81,505
Pacific Northwest	$655	$2,621	$63	$161	$75	$3,039	$6,615

U.S. Department of Transportation (1997); Winston and Langer (2006); Morrison and Winston (2008)

Benefits Forgone

Table 2. The cost of Not building HPPR (millions of USD—2012 present value)

	Transportation Benefits	Economic Output Generated	Tax Revenue Generated	Estimated Project Costs	Net Benefit of HPPR
California	$52,613	$205,200	$23,940	$68,400	$8,153
Chicago Hub	$13,915	$10,200	$1,190	$3,400	$11,705
Northeast Corridor	$81,505	$351,000	$40,950	$117,000	$5,455
Pacific Northwest	$6,615	$25,500	$2,975	$8,500	$1,090
Total 40-year cost of not building HPPR:			$26,403		

U.S. Department of Transportation (1997); California High-Speed Rail Authority (2013); Progressive Railroading (2013); Amtrak (2013); Weisbrod and Reno (2009)

this paper that would be incurred if HPPR is not built, primarily because there are no reliable statistics available at this time to measure these costs or benefits in relation to the construction, maintenance, and operation of HPPR in the United States or abroad.

Furthermore, for the following calculations, additional data was utilized for a minimum top line number for each corridor. Factors that cannot be sufficiently allocated to specific corridors are not included in the bottom line (Tables 1 and 2).

This chapter addresses HPPR – intercity passenger rail service that is highly integrated with other transportation services – as a vital element of the American transportation system. It attempts to address the question: What is the cost of not building an improved intercity passenger rail system as a complementary capacity to the nation's overused, underfunded, and barely-maintained airways and roadways?

Recognizing the limitations of the dominant elements of today's transportation system, and the potential benefits HPPR can provide, policy makers must not forgo the billions of dollars of annual economic, mobility, and energy/environmental benefits as the cost of not building HPPR.

REFERENCES

Airports Council International/North America. (2011). Airport capital development costs 2011 – 2015.

American Association of State Highway and Transportation Officials. (2014). Retrieved from www.highSpeed_rail.org/pages/basicfacts.aspx

American Public Transportation Association. (2012). 10.4 billion trips taken on U.S. public transportation in 2011.

American Public Transportation Association. (2014). Record 10.7 Billion Trips Taken On U.S. Public Transportation In 2013.

American Road & Transportation Builders Association. (2014). Transportation FAQs. Retrieved from www.artba.org/about/transportation-faqs/#20

American Society of Civil Engineers. (2014). 2013 report card for America's infrastructure. Retrieved from www.infrasttructurereportcard.org

Amos, P., Bullock, D., & Sondhi, J. (2010). *High-speed rail: the fast track to economic development, World Bank.* Beijing: R.P., China.

Amtrak. (2012). *Amtrak on pace to set new ridership records.*

Amtrak. (2014). *FY 2015 budget request seeks change in federal passenger rail investment: warns status quo puts Northeast Corridor infrastructure at risk.*

Amtrak website. (2011). Atlanta Constitution. (March 31, 2006). On-time arrival for fifth runway at Atlanta Hartsfield-Jackson Int'l Airport.

Amtrak website. (2014). *Long distance train facts.*

Australian Government, Department of Infrastructure, Transport, Regional Development and Local Government. (2010). *A profile of high-speed railways.*

Ball, M., Barnhart, C., Dresner, M., Hansen, M., Neels, K., Odoni, A., & Zou, B. et al. (2010). *Total delay impact study: a comprehensive assessment of the costs and impacts of flight delay in the United States final report.* National Center of Excellence for Aviation Operations Research.

Baxandall P., Davis B., & Dutzik T. (2012). *Transportation and the new generation: why young people are driving less and what it means for transportation policy.* Frontier Group/U.S. PIRG Education Fund.

Bureau of Transportation Statistics, Research and Innovative Technology Administration. (2011). *Annual report on airline performance.* Washington, D.C.: U.S. Department of Transportation.

De Rus, G. (2008). The economic effects of high-speed rail investment: discussion paper no. 2008-16 revised. Joint Transport Research Centre, University of Las Palmas, Spain.

Dutzik, T., & Steva, E. (2010). *Next stop California: the benefits of high-speed rail around the world and what's in store for California*. California Public Interest Research Group Education Fund.

ENO Transportation Foundation. (2012). *NextGen: aligning costs, benefits and political leadership*. Washington, D.C.

Federal Aviation Administration. (2014), *Fact sheet – FAA forecast – fiscal year 2013 – 2014*. Retrieved from www.gov/news/fact_sheet/news_story.cfm?newsld=15934

Federal Highway Administration. (2009). *National household travel survey*. Washington, D.C.: U.S. Department of Transportation.

Federal Highway Administration. (2013). Status of the nation's highways, bridges, and transit: conditions & performance. *U.S. Department of Transportation,* Washington, D.C., www.fhwa.dot.gov/policy/2013cpr/lb_es.htm

Florida Department of Transportation. (2011, April 15) Airport costs. Retrieved from http//www.Dot.state.fl.us/planning/policy/cost

Government Accountability Office. (2009). *High-speed passenger rail: future development will depend on addressing financial and other challenges and establishing a clear federal role, GAO-09-317.*

Greengauge, (2006), *High-speed trains and the development and regeneration of cities.*

Haas, P. (2014). Modal shift and high-speed rail: a review of the current literature. *Mineta Transportation Institute, Report 12-35.*

Henry, L., & Marsh, D. (2009), Intermodal surface public transport hubs: harnessing synergy for success in America's urban and intercity travel, *Capital Metropolitan Transportation Authority*, Austin, Texas. Retrieved from http://www.amtrak.com/servlet/ContentServer/Page/1248542787937/1237405732517

Jehanno, A., James, C., & Palmer, D. (2011). *High-speed rail and sustainability.* International Union of Railways.

Joint Transportation Research Centre. (2008). *Effects of high-speed rail investment discussion paper no. 2008-16.*

Kageson, P. (2009). Environmental aspects of inter-city passenger transport (Discussion paper no. 2009-28). *Joint Transport Research Centre.*

Langer, A., & Winston, C. (2006). *The effect of government highway spending on road users' congestion costs.*

Machalaba, D. (2007). Crowds heed Amtrak's 'all aboard,' improved service, air, woes lure travelers in northeast; long hauls still suffer. T*he Wall Street Journal.*

Metropolitan Washington Airports Authority. (n. d.). Retrieved from (www.washingtonairports.com/dulles/773

Moijawad, J. (2011), Air service cutbacks hit hardest where recession did. Retrieved from www.newyorktimes.com

Morgan, C., & Sperry, B. (2011). Economic impacts of intercity passenger rail service: evidence from passenger surveys, *Transportation Research Record: Journal of the Transportation Research Board, No. 2261*, Washington, D.C.

Morrison, S., & Winston, C. (2008). The effect of FAA expenditures on air travel delays. *Journal of Urban Economics*, 2008, 63.

Nash, C. (2009). When to invest in high-speed rail links and networks. *OECD/ International Transport Forum*, Madrid, Spain.

Nash, C. (n. d.). High Speed Rail Investment: an overview of the literature. *Institute for Transport Studies, University of Leeds.*

National Surface Transportation Infrastructure Financing Commission. (2008). *Paying Our Way: A New Framework for Transportation.* Washington, D.C.: U.S. Department of Transportation.

Neff, J. (2011). *High-speed rail investment background data.* Washington, D.C.: American Public Transportation Association.

Obama, B. (2011). *State of the Union Address.* Washington, D.C.

Organization for Economic Cooperation and Development. (2001), http://esa.un.org/unsd/sna1993/introduction.asp

Passenger Rail Working Group. (2007). *Vision for the future U.S. intercity passenger rail network through 2050.* National Surface Transportation Policy and Revenue Study Commission.

Reno A. & Weisbrod G., (2009). *Economic Impact of Public Transportation Investment.*

Santini, D., Wang, M., & Warner, S. (1994). *Methods of valuing air pollution and estimated monetary values of air pollutants in various U.S. regions, Argonne National Laboratory*. Argonne, IL: U.S. Department of Energy.

Sivak, M., & Schoet, B. (2011). *Recent changes in the age composition of drivers in 15 countries UMTRI-2011-43*. University of Michigan Transportation Research Institute.

Systra. (2009). *High-speed rail development programme 2008/9, final report for Workstream 2.*

Texas Transportation Institute. (2011). Annual Urban Mobility Report, Texas A&M University.

http://www.progressiverailroading.com/high_speed_rail/article/Midwestern-states-hope- to-take-their-longplanned-highspeed-and-intercity-passenger-rail-network-to-the- next-level--20806

Transportation for America. (2011), *Aging in Place, Stuck without Options: Fixing the Mobility Crisis Threatening the Baby Boom Generation*. Retrieved from http://t4america.org/resources/seniorsmobilitycrisis

U.S. Department of Transportation. (1997). *High-Speed Ground Transportation for America.*

U.S. Department of Transportation. (2012). *2013 Budget.*

Wisconsin Public Interest Research Group Foundation, (2010), *How a Faster Passenger Rail Network Could Speed Travel and Boost the Economy.*

KEY TERMS AND DEFINITIONS

Agglomeration: The act of amassing economic activities in areas, communities and/or corridors.

Airport and/Airway Trust Fund (AATF): The primary funding source for the Federal Aviation Administration, receiving revenues from a series of excise taxes paid by users of the national airspace system. The fund was designed to finance investments in the airport and airway system and, to the extent funds were available, cover the operating costs of the airway system as well. Its revenue sources include excise taxes on: Domestic airline passenger tickets; Domestic airline passenger flight segments; International passenger arrivals and departures; Air cargo waybills; Aviation fuels; and Amounts paid for the right to provide mileage awards.

American Recovery and Reinvestment Act (ARRA): Signed into law by President Obama on February 17th, 2009, ARRA was an unprecedented effort to jumpstart the nation's economy, create or save millions of jobs, and make a down payment on addressing long-neglected challenges so the U.S. can thrive in the 21st century. The act authorized $8 billion toward the development of a network of what the administration called high-speed rail passenger service intended to evolve over time to rival the high-speed rail service currently experienced in Europe, Japan and other regions of the world.

Amtrak: The trademark name for the National Passenger Railroad Corporation.

DOT: The U.S. Department of Transportation.

Federal Aviation Administration (FAA): Is the modal agency within the U.S. Department of Transportation responsible for administering the operation and safety of the nation's commercial and private aircraft aviation system. The FAA is funded primarily through the Airports and Airway Trust Fund and through general revenue appropriated by the U.S. Congress.

Federal Highway Administration (FHWA): Is the modal agency within the U.S. Department of Transportation responsible for coordinating the development and maintenance of the nation's national highway system. The FHWA is funded primarily through the Highway Trust Fund and more recently from general revenue appropriated by the U.S. Congress.

Federal Railroad Administration (FRA): Is the modal agency within the U.S. Department of Transportation responsible for maintaining the safety of the nation's freight and passenger rail network, and, as a result of ARRA and subsequent congressional legislative authorizations and appropriations, a variety of grant and loan programs to support the development and reconstruction of the nation's freight and passenger rail network.

Federal Transportation Administration (FTA): Is the modal agency within the U.S. Department of Transportation responsible for administering grant, planning and safety programs for the nation's bus and rail transit organizations. The agency is funded in part from the Highway Trust Fund and in part from general revenue appropriated by Congress.

Higher Performing Passenger Rail (HPPR): Is the newly introduced terminology for the administration's previously announced high-speed passenger rail initiative. Its components include: Core Express Service: Frequent trains at 125-250+mph in the nation's most dense and most populous regions; Regional Service: Operating at 90-125mph between mid-sized and large cities; and Emerging Service: Operating at speeds up to 90mph, connecting communities to the passenger rail networks and providing a foundation for future corridor development.

High-Speed Rail (HSR): Is defined by the International Union of Railways as specially built passenger rail lines equipped for speeds generally equal to or greater than 250 km/h; specially upgraded to handle speeds of the order of 200 km/h with special features that address topographical, relief or town-planning constraints, on which the speed must be adapted.

Highway Trust Fund: The source of revenue for the interstate highway system and other federal-aid highway programs, was created by the Highway Revenue Act of 1956. The trust fund has been reauthorized periodically with the current reauthorization will expiring in 2015. The fund has two accounts, a highway account that funds the federal aid highway program and other FHWA programs, and a transit account that funds a variety of programs through the Federal Transit Administration.

Induced Travel: An increase in traffic volume that occurs soon after new transportation options become available to potential travelers who have not used, or have been precluded from using previously available transportation options.

Mode Shift: Travelers using transportation modes different from their usual mode choice.

NextGen: An umbrella term for the ongoing transformation of the National Airspace System (NAS). At its most basic level, NextGen represents an evolution from a ground-based system of air traffic control to a satellite-based system of air traffic management.

Passenger Rail Corridors: 11 rail corridors identified in the 1991 Intermodal Safety Transportation Efficiency Act (ISTEA) to receive attention due to the high number of highway/rail conflicts. By their recognition they were designated "high-speed corridors." These corridors included the Pacific Northwest, California, the Chicago Hub, South Central, the Gulf Coast, Florida, Southeast, Northeast, Keystone, Empire, and Northern New England.

Passenger Rail Investment and Improvement Act (PRIIA) of 2008: Reauthorizes the National Railroad Passenger Corporation, better known as Amtrak, and strengthens the U.S. passenger rail network by tasking Amtrak, the U.S. Department of Transportation (US DOT), Federal Railroad Administration (FRA), states, and other stakeholders in improving service, operations, and facilities. PRIIA focuses on intercity passenger rail, including Amtrak's long-distance routes and the Northeast Corridor (NEC), state-sponsored corridors throughout the nation, and the development of high-speed rail corridors.

Passenger Rail Working Group (PRWG): Was created by the National Surface Transportation Policy and Revenue Study Commission, which was authorized by 109th Congress in Section 1909 of the Safe, Accountable, Flexible, Efficient, Transportation Equity Act – A Legacy for Users (SAFETEA-LU) to provide Congress with a national surface transportation vision, with supporting funding and policy recommendations to preserve and enhance the surface transportation system of the United States for the next 50 years. The PRWG was charged with providing to the Commission: recommendations on a national vision for intercity passenger rail through 2050; a cost estimate for that vision; a federal funding program for passenger rail; and a governance structure for program development. Members of the working group included: Frank Busalacchi, Wisconsin Department of Transportation and member of the National Surface Transportation Policy and Revenue Study Commission; John Bennett, Amtrak; Ross Capon, National Association of Railroad Passengers; Larry Ehl, Washington Department of Transportation; David Ewing, States for Passenger Rail Coalition; Mehdi Morshed, California High-Speed Rail Authority; Patrick Simmons, North Carolina Department of Transportation; and Eugene Skoropowski, Capitol Corridor Joint Powers Authority.

Chapter 5
Emerging Challenges and Opportunities for U.S. High-Speed Rail Development

Francis P. Banko
WSP | Parsons Brinckerhoff, USA

Jackson H. Xue
WSP | Parsons Brinckerhoff, USA

ABSTRACT

As we witness the advancement of U.S. high-speed rail initiatives, the country can look towards its European and Asian counterparts for best practices and lessons learned from their decades of high-speed rail design and operations. These experiences gained may be applicable towards projects such as the Texas Central Railway and the California High-Speed Rail Project. This chapter will address the events of 2009 that have brought domestic high-speed rail to the forefront of U.S. rail transportation. This includes the new FRA Tier I and proposed Tier III criteria, challenges associated with each FRA tier of operation, overseas interoperability efforts, snapshots of international experiences (from policy and technological perspectives), the holistic system-based approach to safety, ongoing efforts of the FRA Engineering Task Force, and additional challenges and opportunities moving forward.

DOI: 10.4018/978-1-5225-0102-2.ch005

INTRODUCTION

Regardless of operating speed, it is essential that trains are safely and reliably operated if train travel is to be a preferred mode of transportation. With its rich rail history, the U.S. has had substantial experience in developing and enforcing safety requirements for passenger rail equipment and systems. Title 49, Parts 200-299 of the Code of Federal Regulations (CFR) identifies criteria established by the Federal Railroad Administration (FRA) to govern the design and operation of passenger and freight rail rolling stock. However, these criteria are only applicable for equipment that operate up to 125 mph (201 km/h), also known as Tier I operation, or 150 mph (241 km/h), also known as Tier II operation. To date, there are no established U.S. regulations or standards that address design of rail equipment to operate at speeds greater than 150 mph (241 km/h).

To facilitate introduction of service at speeds greater than 150 mph (241 km/h), the U.S. has looked to its overseas counterparts for best practices from their decades of experience with the design, manufacture, operation, and maintenance of high-speed rail (HSR) equipment and systems. The information gathered though interaction with HSR industry experts will aid the development and subsequent growth of HSR policies and technologies in the U.S. This chapter provides insight into the challenges faced when introducing new HSR technologies and associated design and construction techniques, along with the challenges of complying with current U.S. regulations. These issues are being overcome as the domestic rail industry looks to international high-speed railroads, and applies the lessons learned by those who have been successfully providing safe, reliable, and efficient HSR service.

This chapter addresses the following elements:

- **History of HSR Support:** In the U.S., HSR is "old, but new." Ever since the introduction of passenger rail service in the U.S., there have been numerous examples where speed records were chased and surpassed. More recently, one can look to Amtrak's Acela Express service, where trainsets are in daily operation on the Northeast Corridor (NEC) at speeds up to 150 mph (241 km/h). The American Recovery and Reinvestment Act of 2009 (ARRA) has since served as the catalyst for new investments in HSR in existing and new corridors throughout the U.S.
- **Challenges and Considerations from the Project and Industry Perspectives:** New HSR initiatives in the U.S. are focused largely on two different maximum operating speeds (Tier I and proposed Tier III operations up to 220 mph (354 km/h)). Each of these speed ranges introduces a unique set of challenges, including: challenges with traveling at these speeds, challenges

with implementing existing regulations and statutory requirements, and challenges with introducing new technologies.

- **Overseas Experiences:** "How do they do it over there?" Recognizing the decades-long safe and reliable operation of HSR throughout Asia and Europe, this section introduces international best practices associated with the planning, design, manufacture, testing, commissioning, and operation and maintenance of HSR trainsets and systems. The concept of a holistic system-based approach to HSR, and the inherent improvements to safety and reliability that are realized when such an approach is implemented (a common theme worldwide), is a critical attribute towards providing safe, reliable, and efficient service.
- **Leveraging International Best Practices and Lessons Learned:** NOT re-inventing the wheel! The U.S. is in an enviable position to leverage lessons learned, and to adopt service-proven methodologies to successfully deliver its HSR initiatives. These methodologies will be introduced, highlighting the ongoing work of the FRA's Engineering Task Force (ETF), and the FRA's development of new regulations that provide the framework for alternatively-designed passenger rail equipment for use in Tier I service and that will support new Tier III operations.
- **Challenges Moving Forward:** Even with the excellent progress made by the FRA in developing regulations that will support the implementation of international service-proven HSR methodologies, additional challenges remain. Examples of these challenges will be introduced, which include those pertaining to static axle loads and trainset weight, vehicle track interaction for higher speed operations, and trainset aerodynamics and the associated effects on exterior noise.
- **Opportunities:** This section will highlight the technological and operational opportunities associated with developing HSR in the U.S. For example, in Europe, a movement towards interoperability of the HSR network is underway. U.S. initiatives that focus on developing a national HSR network can benefit from Europe's lessons learned. Technological elements and advances that contribute to the environmentally "green" attributes of HSR including power regeneration, lightweight trainset designs, and optimized aerodynamics will be introduced, as well as opportunities associated with the development of domestic HSR manufacturing capabilities.

History of HSR Support

In the U.S., HSR is "old, but new." The quest to increase the speed of rail transportation flourished in the U.S. during the golden age of rail, with speed records being conquered by steam locomotives (such as the Empire State Express, in 1893, between Buffalo and New York City, which reached 112.5 mph (181 km/h) (Roberts, 2013)), and with even higher speed records being achieved for gas turbine powered locomotives in the 1960's.

The most practical example of U.S. HSR is Amtrak's Acela Express service, inaugurated in 2000, with electric trainsets in daily operation on the NEC at speeds up to 150 mph (241 km/h). ARRA has since served as the catalyst for new investments in HSR in existing and new corridors throughout the U.S. For the future, programs like the Texas Central Railway and the California High-Speed Rail Project are envisioning revenue operations at speeds up to 205 mph (330 km/h) and 220 mph (354 km/h), respectively, representing new milestones for the U.S.

U.S. HSR initiatives were reinvigorated when President Obama introduced his administration's blueprint for a new high-speed passenger rail network in America. "What we need…is a smart transportation system equal to the needs of the 21st century," the President stated, "a system that reduces travel times and increases mobility, a system that reduces congestion and boosts productivity, a system that reduces destructive emissions and creates jobs " (Knowlton, 2009). The President's *Vision for High-Speed Rail in America* supported four strategic transportation goals: ensuring safe and efficient transportation choices, building a foundation for economic competitiveness, promoting energy efficiency and environmental quality, and supporting interconnected livable communities (FRA, 2009).

Ten potential HSR corridors were identified for development (refer to Figure 1): Pacific Northwest, California, South Central, Gulf Coast, Florida, Southeast, Chicago Hub Network, Empire, Keystone, and Northern New England. Funding would be available through the ARRA Capital Assistance for High-Speed Rail Corridors and Intercity Passenger Rail Service program, from a pool of $8 billion intended to be the catalyst for establishing a world-class passenger rail system. In addition, the existing NEC, which supports America's only HSR service, would be permitted to compete for funds. An additional $1.3 billion would be available in the form of capital grants to the National Railroad Passenger Corporation (Amtrak). Currently, the ARRA program is supporting various HSR initiatives throughout the U.S., where existing rail corridors are being upgraded for 125 mph (201 km/h) operation, and new, dedicated corridors are being planned and designed for the higher operating speeds.

In November 2009, the FRA Office of Railroad Safety and Office of Railroad Development published the document entitled *High Speed Passenger Rail Safety Strategy*, wherein the agency committed to "take on the important work of helping to make [HSR] a reality in markets across the Nation" (FRA, 2009). Per FRA, "the hallmark of world-class, [HSR] is safety" (FRA, 2009) and as such, developed a

Figure 1. Candidate U.S. HSR corridors (FRA, 2009)

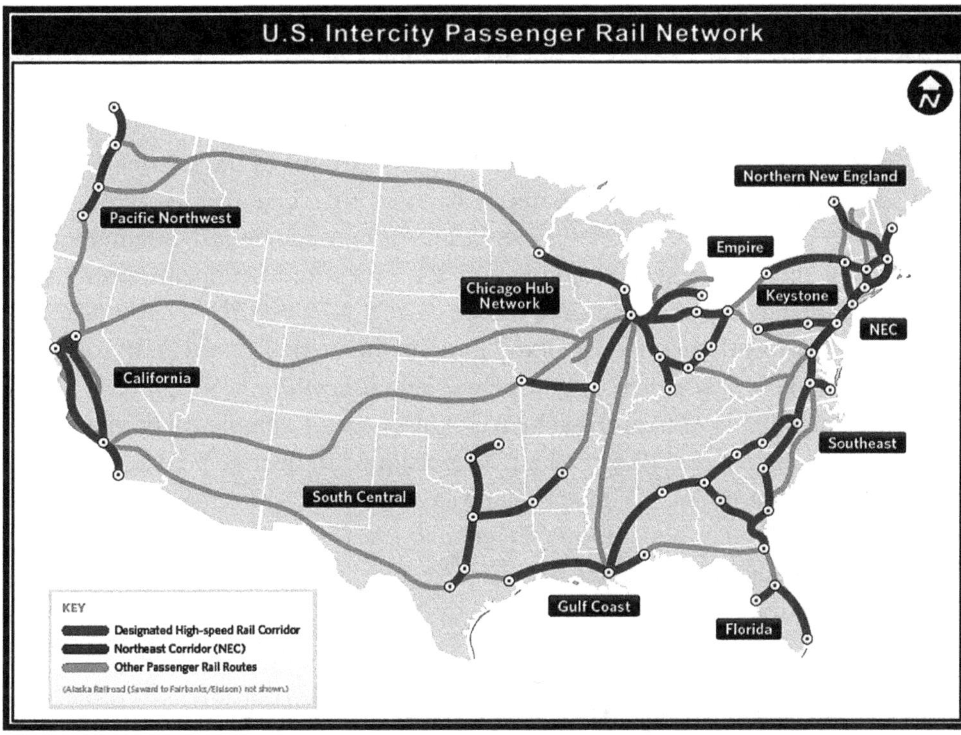

three-tiered approach which includes: establishment of HSR safety standards and program guidelines, application of a system-safety approach to safety concerns, and assurance that train emergencies can be effectively and efficiently managed by passenger railroads (FRA, 2009).

Appendix B of the *High Speed Passenger Rail Safety Strategy* initially identified a Tier structure consisting of Tier 0 through Tier V, as based on maximum operational speeds. As part of the efforts to develop new safety standards and regulatory text to accommodate service-proven equipment, the Tier structure has been simplified to include three tiers[1]:

Tier I – defined in 49 CFR §238.5 as "operating at speeds not exceeding 125 mph [(201 km/h)]." Tier I examples include the current New Jersey Transit's, Long Island Rail Road's, and Metro-North Railroad's commuter rail operations in the Northeast and Caltrain's and Metrolink's commuter rail operations in California. New Tier I operations being planned include the Chicago to St. Louis Corridor, a 284 mile (457 km) long line with initial operating speeds up to 110 mph (177 km/h); the All Aboard Florida railroad, a proposed line linking Miami to Orlando at speeds up to 125 mph (201 km/h); and the Pacific Northwest Rail Corridor, a proposed line between Eugene, OR and Vancouver, BC, with a maximum speed of 110 mph (177 km/h).

Tier II – defined in 49 CFR §238.5 as "operating at speeds exceeding 125 mph (201 km/h), but not exceeding 150 mph (241 km/h)." The only service in the U.S. currently operating in a Tier II environment is the Amtrak Acela Express, where 150 mph (241 km/h)[2] operation is attained on two sections of track – one in Rhode Island and the other in Massachusetts. It is noted that Tier II is currently being evaluated to cover up to 160 mph (258 km/h). Proposed Tier III rulemaking accommodates this higher speed regime, with the exception that operations must be within a dedicated right-of-way.

Tier III – currently being considered in draft rule text as speeds not exceeding 125 mph (201 km/h) in a shared right-of-way and exceeding 125 mph (201 km/h), but not 220 mph (354 km/h) in an exclusive right-of-way. Examples of proposed Tier III services include XpressWest, which is envisioned to operate between southern California and Las Vegas, NV at speeds up to 150 mph (241 km/h); and California High-Speed Rail, where operations are envisioned at speeds up to 220 mph (354 km/h) between San Francisco, CA and Los Angeles, CA, with future extensions to Sacramento and San Diego.

CHALLENGES AND CONSIDERATIONS FROM THE PROJECT AND INDUSTRY PERSPECTIVES

New HSR initiatives in the U.S. are focused largely on two different maximum operating speeds - Tier I and the proposed Tier III. Each of these speed ranges presents a unique set of challenges implementing existing regulations and statutory requirements, while introducing new technologies to meet the performance requirements of the specific project.

Tier I Challenges and Considerations

HSR projects operating in the speed range categorized by the FRA as Tier I can implement a broad range of rail vehicle and core system technologies. Key core systems include signal and train control, communications, traction power, and infrastructure. Projects must comply with the signal and train control requirements, including the requirements of a Positive Train Control (PTC) system, as identified in 49 CFR § 236 Subpart I. From a project perspective, implementing PTC can be a challenge when a HSR system is operating over track that is shared by other passenger and/or freight operations, as all equipment needs to be PTC compliant. From a technological perspective, the industry is still maturing as PTC systems and approaches are being developed, and requests for FRA approval are being considered. Projects also need to consider the regulatory requirements for communication systems, such as those contained in 49 CFR § 238; specific system requirements, as dictated by the physical and operational environments; and emergency preparedness requirements. Decisions on whether the project will be electrified using an overhead contact system (OCS), for example, or whether diesel-electric locomotives will be utilized will determine environmental impacts and will dictate requirements of a traction power supply system. Lastly, the project must comply with minimum requirements for track, based on the maximum allowable operating speed, as defined by the FRA's track classes (refer to Table 1).

Depending on the operational requirements of the system, there are numerous rail vehicle technologies that can be considered. Diesel-electric and electric locomotives can haul single-level and multi-level coaches in a bi-directional push-pull configuration. Multi-level coaches will carry more passengers than a single-level coach of similar length. Passenger ingress to and egress from a coach is typically faster on a single-level coach as choke points (e.g., stairs, transitions between levels) are avoided. Electric or diesel multiple unit (E/DMU) trainsets (where traction power equipment is distributed throughout the trainset) are available options. An EMU or DMU trainset configuration will carry more passengers than a locomotive-hauled trainset of similar length and configuration, as passengers are carried on all units of the trainset. When compared to locomotive-hauled trains, EMU/DMU trainsets may have a higher initial capital cost per passenger seat, and require different maintenance programs. There are also dual mode variants whereby electric

Table 1. 49 CFR § 213 classes of track

Class of Track	Maximum Operating Speed Freight, mph (km/h)	Maximum Operating Speed Passenger Trains, mph (km/h)
Excepted Track	10 (16)	N/A
Class 1 Track	10 (16)	15 (24)
Class 2 Track	25 (40)	30 (48)
Class 3 Track	40 (64)	60 (97)
Class 4 Track	60 (97)	80 (129)
Class 5 Track	80 (129)	90 (145)
Class 6 Track	*	110 (177)
Class 7 Track	*	125 (201)
Class 8 Track	*	160 (258)
Class 9 Track	*	220 (354)

* Freight may be transported at passenger train speeds with certain conditions being met.

and diesel motive power options can be provided on one trainset, as well as options to draw power from multiple sources (e.g., OCS and third rail, and multiple voltages). Dual mode variants will be more costly than single mode designs; however, greater operational flexibility is provided (e.g., passengers are provided a one seat ride and locomotives do not need to be changed when traveling between electrified and non-electrified portions of the railroad).

The challenge lies in selecting the best rail vehicle technology for the system, based on the system's unique characteristics and the requirements of the operating plan. The selection of particular rail equipment is influenced by several factors. There needs to be an understanding of operational and commercial requirements, inclusive of ridership projections and travel time. The projections and travel time will factor into the maximum operating speed required; diesel-electric locomotives capable of operating up to 125 mph (201 km/h) are being developed whereas several 125 mph (201 km/h) capable electric locomotive designs are in service. Funding availability will play into whether or not electrification of the railroad is possible; electrification of a railroad requires additional right-of-way to accommodate the traction power and OCS equipment, and utility connections. Coordination is required between existing railroad infrastructure and core systems (e.g., electrification, stations and platforms, maximum authorized speed), existing rail vehicle technology already in use (which can allow for economies of scale when procuring additional equipment and spare parts), and existing maintenance facilities (which were designed to accommodate a specific type of rail vehicle technology). Also, consideration should be given towards environmental impacts; diesel-electric locomotives are required to comply with the Environmental Protection Agency (EPA) emissions standards, as codified in 40 CFR §1033, and may produce higher levels of exterior noise and vibration when compared to equivalent electric locomotives.

Tier II Challenges and Considerations

Amtrak's Acela Express service represents the single example of FRA Tier II HSR operation, with speeds exceeding 125 mph (201 km/h) and up to 150 mph (241 km/h). FRA's Tier II requirements were developed to support Amtrak's HSR program whereby trainsets would operate at speeds in excess of the Tier I regulations on track shared by other passenger and freight rail equipment. An element of the Tier II regulations is the prohibition on occupied lead cars; the regulation was developed around the premise of a concentrated power trainset configuration, where the leading and trailing ends of the trainset provide the motive power for the trainset, as opposed to an EMU trainset design. The regulations include requirements for crash energy management (CEM) systems to be installed to dissipate kinetic energy during a collision, and structural requirements for the underframe of the cab of a power car to resist a minimum longitudinal static compressive force of 2,100,000 lb_f (9341 kN) without permanent deformation to the cab. The Tier II requirements resulted in industry designing and manufacturing a purpose built trainset, the Acela, as opposed to industry supplying a trainset that was already proven in HSR service overseas. Recently, Amtrak attempted to procure additional Acela coaches; however, due to the unique attributes of the trainset design, and the limited market for the Acela, this procurement was cancelled in favor of procuring Tier III-compliant trainsets.

Tier III Challenges and Considerations

Learning from the Tier II regulatory process, the FRA and industry recognized the need to develop a new set of regulations (Tier III) that would apply to HSR operations with speeds not exceeding 125 mph (201 km/h) in a shared right-of-way and exceeding 125 mph (201 km/h), but not 220 mph (354 km/h) in an exclusive right-of-way. The challenge was to develop regulations that would leverage the benefits of existing, service-proven HSR trainset design and operating practices, while maintaining a level of safety equivalent to that provided by the current CFR. These regulations, as they are currently being developed, define requirements for trainset structure and crashworthiness, glazing, braking, interior fittings and surfaces, emergency systems, cab equipment, fire safety, safety appliances, inspection, testing, and maintenance. The requirements were developed based on the lessons learned and best practices from international HSR operations, and will allow industry to supply Tier III-compliant trainsets to U.S. projects whereby much of the existing, service-proven HSR trainset design is preserved.

Other Challenges and Considerations

Additional challenges exist – one of which is associated with the procurement of HSR trainsets when Federal funds are expended on a project, where there is a requirement that the trainsets be compliant with FRA's Buy America regulation. FRA's Buy America statute, 49 U.S.C. § 24405(a) and applicable FRA guidance[3], allows the U.S. Secretary of Transportation to obligate funds for "a project only if the steel, iron, and manufactured goods used in the project are produced in the United States." FRA has stated that what constitutes FRA Buy America compliant rolling stock is rolling stock that have undergone final assembly in the U.S. from components that are manufactured in the U.S. Currently, there is no domestic HSR trainset industry; therefore, the Buy America requirement represents the challenge of developing sustainable domestic HSR manufacturing and assembly capabilities that result in effective transfers of technology. When successfully realized, the outcome will be the development of a U.S. HSR industry and the creation of high quality jobs.

Another significant challenge lies in the correct integration of all elements of a HSR system. The following section introduces the concept of a holistic system-based approach to implementing HSR – an approach that is based on lessons learned from international counterparts.

OVERSEAS EXPERIENCES

How do they do it over there? As U.S. HSR corridors are being developed, there is an opportunity to investigate international HSR operations. Understanding European and Asian best practices associated with planning, designing, manufacturing, testing, commissioning, and operation and maintenance of HSR trainsets and systems provides a solid framework from which U.S. HSR initiatives can benefit. By learning answers to the question "How do they do it over there?" one can distill from those answers the applicable best practices.

Introduction of HSR Operation Worldwide

There are numerous HSR operations worldwide that are suitable candidates for this investigation. One can look to Germany, Korea, Japan, China, Spain, and France for cross-sections of safe, reliable, and efficient HSR operations. In particular, focus is made on those operations that are or will be capable of providing service at speeds of up to 220 mph (354 km/h).

Germany: Deutsche Bahn (DB)

DB's HSR operations began in 1991 with the opening of the high-speed line between Hamburg and Munich. By 1998, the high-speed network had expanded to Cologne. In 2003, a high-speed line was established between Cologne and Frankfurt with trains operating at a maximum speed of 186 mph (300 km/h). In 2006, DB upgraded the infrastructure between Cologne and Berlin to facilitate increases in operating speeds.

Korea: Korea Railroad Corporation (Korail)

The Korean National Railroad (KNR) was in charge of railroad operations in South Korea since 1963. In 2005, KNR was divided into two agencies: Korea Rail Network Authority, which is responsible for railroad construction and management, and Korail, which is responsible for railroad operations. Korail currently operates HSR between the cities of Seoul and Busan (Gyeongbu HSR) up to a maximum speed of 190 mph

(305 km/h). High-speed lines between Seoul and Mokpo (Honam HSR) and Suseo to the existing Gyeongbu HSR will eventually see 218 mph (350 km/h) operations.

Japan: Ministry of Land, Infrastructure, Transport and Tourism (MLIT)

Japan began developing the world's first HSR system in 1964, which operated from Tokyo to Shin-Osaka; this line was named the Tokaido Shinkansen. Japan has over 1,300 miles (2,092 km) of HSR in operation, approximately 400 miles (644 km) under construction, and over 300 miles (483 km) in the planning stage.

Japan's current HSR network comprises eight lines:

- The Tokaido Shinkansen, which opened in 1964. The maximum speed on the Tokaido Shinkansen has increased from 137 mph (220 km/h) to 168 mph (270 km/h);
- The Sanyo Shinkansen, which opened in 1972 and operated from Shin-Osaka to Okayama and was then extended to Hakata in 1975;
- The Tohoku Shinkansen, which opened in 1982, with a maximum operating speed of 186 mph (300 km/h). The maximum operating speed on this line is now 199 mph (320 km/h);
- The Joetsu Shinkansen, which opened in 1982;
- The Akita Shinkansen, which opened in 1997;
- The Yamagata Shinkansen, which opened in 1999;
- The Hokuriku Shinkansen, which opened in 1997; and
- The Kyushu Shinkansen, which opened in 2004.

There are plans to extend the Shinkansen network, with the new Hokkaido Shinkansen currently under construction to bring high-speed trains north to Sapporo.

China: Ministry of Railways of the People's Republic of China (MOR)

China's Ministry of Railways (MOR) was responsible for passenger services, regulation of the rail industry, and development of the rail network and rail infrastructure. In 2013, MOR was dissolved and MOR's responsibilities were divided and assigned to three entities. The Ministry of Transport is responsible for administrative functions including safety and regulations. The China Railway Corporation is responsible for commercial functions including construction, management, railway dispatch, and command. The State Railways Administration is responsible for developing standards, supervising railway work safety, and auditing the quality of transport service and railway projects.

In 2008, China began operation of its first high-speed line between Beijing and Tianjin. Since then, over 6,800 miles (11,000 km) of HSR, with sections capable of 218 mph (350 km/h) operation, have entered into service.

Spain: Renfe Operadora (Renfe)

In 1941, the Red Nacional de los Ferrocarriles Españoles (Renfe) was formed with the nationalization of Spain's railways. In 2005, Renfe was divided into two agencies: Administrador de Infraestructuras Ferroviarias (ADIF), which is in charge of rail infrastructure, and Renfe Operadora, which is in charge of railroad operations (intercity, commuter, and freight). In Spain, there are over 1,900 miles (3,100 km) of HSR operating at speeds up to 193 mph (310 km/h).

France: Société Nationale des Chemins de fer Français (SNCF)

SNCF operates several HSR lines domestically, as well as HSR lines in the United Kingdom, Spain, Morocco, Taiwan, Italy, Belgium, Netherlands, and Portugal. Its domestic HSR lines, also known as Lignes à Grande Vitesse (LGV), operate at speeds up to 199 mph (320 km/h), and include:

- East Line (LGV Est européenne), between Paris and Baudrecourt;
- South-East Line (LGV Sud-Est), between Paris and Lyon;
- Atlantic Line (LGV Atlantique), between Paris and Le Mans and Tours;
- Rhône-Alpes Line (LGV Rhône-Alpes), between Lyon and Valence;
- North Line (LGV Nord), between Paris, Lille, and the Channel Tunnel;
- Mediterranean Line (LGV Méditerranée), between Valence and Marseille; and
- Rhin-Rhône Line (LGV Rhin-Rhône), between Dijon and Mulhouse, Dijon and Lyon, and joining the LGV Sud-Est Line.

Introduction of European and Asian Best Practices

A strategy implemented by HSR operators is the implementation of a holistic system-based approach to delivering HSR. It has been common practice in the U.S. to focus on individual elements of a rail system (e.g., infrastructure, rolling stock, traction power, signals, communications) with each of these elements being developed against a set of independent regulations, standards, and specifications. International HSR operators have recognized that each element of the system plays a role in delivering service, and those benefits (e.g., improvements in safety, reliability, and efficiency) can be realized when the elements of a HSR system are developed in an integrated fashion. This integrated system-based approach is illustrated in the specifications, standards, and codes developed in Europe, Japan, and China.

Europe's Technical Specifications for Interoperability

The European Railway Agency (ERA) has drafted a compendium of *Technical Specifications for Interoperability* (TSI) that identify minimum requirements for each element of a HSR system. Each section of the TSI is integrated at the system level and includes sections on Infrastructure, Energy, Rolling Stock, Operations and Traffic Management, as well as transversal specifications developed for Control Command and Signaling, Telematic, Safety in Railway Tunnels, and Persons with Reduced Mobility, the European equivalent to the U.S. requirements stemming from the Americans with Disabilities Act (ADA). The anticipated outcome of adopting the requirements of the TSI is to ensure the interoperability of the trans-European high-speed and conventional rail systems.

Technical Regulatory Standards on Japanese Railways

Japan's MLIT has compiled an integrated set of standards titled *Technical Regulatory Standards on Japanese Railways*. These standards identify requirements relating to Staff (e.g., operations, training); Guide Way; Stations and Halt (e.g., signal station or marshaling yard); Intersection with Road, Electric Facilities, Operation Safety Facilities (e.g., signaling and communication facilities); Rolling Stock; Other Facilities (e.g., for recording the operating condition of trains); Maintenance Facilities and Rolling Stock; Train Operation; and Special Railways. These standards are intended to provide minimum requirements that when implemented will result in a safe operation. Since its inception in 1964, the Shinkansen has achieved a flawless safety record with no passenger fatalities; to date, the only incidents encountered on the Shinkansen have been several derailments due to earthquakes.

China Code for Design of High-Speed Railway

In 2009, the now dissolved MOR developed design standards for HSR titled *Code for Design of High-Speed Railway*. This code is comprised of numerous chapters including: Overall Design, Transport Organization, Alignment, Earth Structure, Bridge and Culvert, Tunnel, Track, Station and Yard, Traction Power Supply, Electric Power, Communication, Signaling, Information, Disaster Prevention and Safety Monitoring, EMU Facilities, Comprehensive Maintenance, Water Supply and

Sewage, Building Construction, Integrated Earthing, and Environmental Protection. This code was developed with the intent to unify HSR technical standards to support the safe operation of HSR in China.

Each of the documents cited above follows the integrated system-based approach previously introduced and further described below.

International Perspectives on Safety

A Snapshot of HSR Safety and Experiences

The safety records of the European and Asian HSR systems are outstanding with respect to passenger safety, making travel by high-speed train among the safest modes of passenger transportation in the world. A key strategy to delivering safe HSR is through the implementation of a system-based approach to safety that embraces the tenets of fail-safe train separation, CEM technology, and the development of safe and effective procedures for the operation of the railroad, including the efficient implementation of these procedures by a well-trained staff.

China

MOR representatives have advised that the purpose for train operation is the same in China as it is in the U.S. – that is, provide safety and comfort to the passengers and allow them to reach their destination. China's HSR system is constructed under the guidance and regulations of the Ministry.

Every process from the planning, surveying, manufacturing, and operating stages are recommended to be monitored. Newly built track and existing track are interoperable, and dedicated passenger lines are provided for operations above 155 mph (250 km/h). In addition, China utilizes dedicated trainsets for specific lines (e.g., trainsets with a maximum speed of 155 mph (250 km/h) will only be permitted to travel on 155 mph (250 km/h) lines). On existing lines where high-speed trains share track with freight equipment, the maximum speed is limited to 124 mph (200 km/h); this speed is only permitted on lines that have been upgraded.

China is focused on safety and reliability, as well as passenger and crew comfort. Infrastructure requires less maintenance because of the use of slab track (ballast-less track construction consisting of a continuous slab of concrete with the rails supported directly on the slab). A reduced need for maintenance relates to a safer environment.

Operating safe trains depends not only on the operator, but also on the train control design, the trainset design, and the infrastructure design. The concept implemented in China was to build elevated lines to eliminate at-grade crossings; according to China's statistics, most accidents occur at grade crossings. Further, all tracks are fenced to protect against pedestrians accidentally entering the right-of-way; lines with speeds greater than 75 mph (120 km/h) are constructed with controlled access. It is equally important to have and maintain good track quality, and that track quality needs to be closely monitored during manufacturing and while in-service.

China has advised further that manufacturing, operations, and maintenance must be well organized. Even though there are multiple types of HSR equipment in operation, procedures for maintenance have been unified and strict regulations and rules are in place to ensure the quality of maintenance and safety.

France

In Europe, trains are one of the safest modes of transportation. Statistics on fatality rates by mode of transport and types of rail accidents in Europe are illustrated in Figures 2 and 3:

HSR is a complex system (i.e., operation in an open environment, impacts from human factors). Risks exist, therefore, and must be managed via active safety (part of the railway system), passive safety (part of the trainset), and the integration of

Figure 2. Rates of the three primary modes of travel showing factors of increase compared with train travel (Lacôte, 2010)

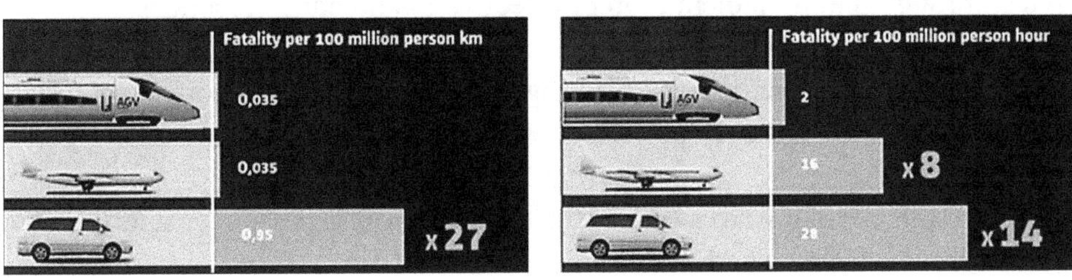

Figure 3. Types of rail accidents (Lacôte, 2010)

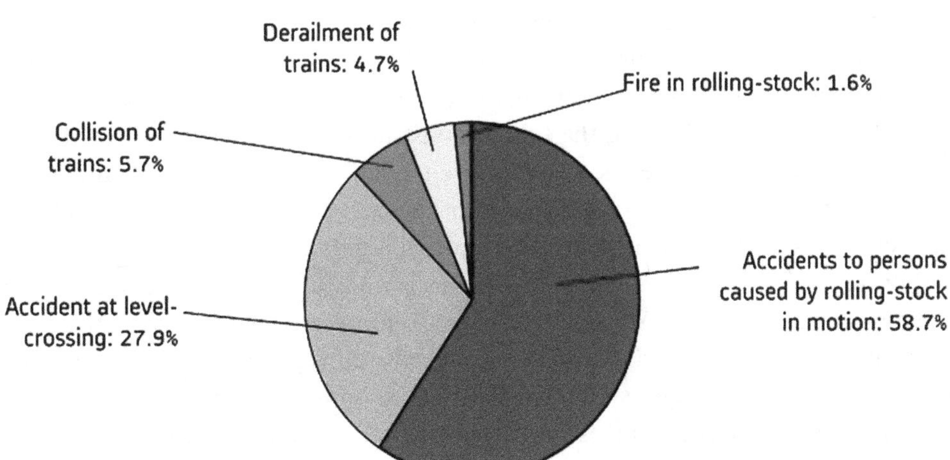

standards as well as by having safety assurance[4] and a safety management system[5] in place to effectively manage potential rolling stock incidents.

It is essential to first prevent accidents by doing all that is possible within reasonable limits and within the limits of existing technologies. For example, all high-speed lines in France are fenced and detectors are installed to sense falling objects (from overpasses) via nets. These nets automatically signal the automatic train control (ATC) system to stop the train if an object is detected. Bogie and track monitoring is performed to ensure safe high-speed travel. Wayside protection includes detection of broken rails by track circuits, switch point control, crosswind speed monitoring, earthquake detection, and overspeed protection and control. Passive safety must also be considered to limit the consequences of potential accidents. Elements of passive safety include, for example, trainset crashworthiness and fire safety.

The TSI was developed primarily for interoperability and the possibility of maintaining an open market for the procurement of trainsets and core systems. While European norms (Euro Norms) are not requirements or regulations that must be followed (unless specifically specified), their marriage with the TSI provides for a safe system. Safety is not only relevant prior to service, but is also essential during service, through the implementation of effective maintenance practices.

When developing a new HSR system, it is preferable to have all entities involved in the early phases of the design of rolling stock, infrastructure, core systems, etc. The implementation of a system-based approach is absolutely necessary.

Germany

HSR experts in Germany advised that the key to crew and passenger safety is compliance with the TSI and national standards.

DB uses a hazard analysis process to assess the level of risk/hazard severity associated with a prescribed (e.g., 164-feet (50-m)) section of infrastructure. For example, if the track runs at-grade, a lower risk is assigned; conversely, if the track runs on an aerial structure, a higher level of risk is assigned. DB also analyzes the risk curve for each particular line and vehicle and determines the operational characteristics appropriate for that line. For example, in areas of high crosswinds, DB reduces the speed limit of the line or constructs wind barriers. Right-of-ways are fenced in with controlled access to prevent entry of trespassers and animals. In the end, the risk must be within the limit of the curve developed. The operational characteristics are documented in DB's internal regulations.

For new HSR endeavors, DB highly recommends using ballast-less track construction (i.e., slab track). Although slab track is more expensive upfront, operators prefer it over ballasted track construction because of reduced costs for inspection and maintenance. Another aspect that DB is keen on is the protection of crew and passengers against rapid changes in pressure caused when a train enters and exits a tunnel. To that end, it is essential to have a good trainset sealing system incorporated into the design of the train.

Spain

Renfe places a high level of importance on safety and security, and advised that it is looking to install security cameras inside and outside of the trainsets to monitor crime, thefts, terrorist activities, etc. The Spanish operator recommended that cameras installed on the trainsets provide three types of surveillance whereby the interior of the train is monitored, the pictures of passengers boarding the train are recorded, and the cameras are protected from vandals looking to disable them.

Holistic System-Based Approach to Safety

The overall safety of a HSR system is assured though the individual contributions to safety provided by each subsystem. The correct integration of each subsystem into a system provides assurance that the individual subsystem elements of safety are preserved. This section provides several examples of how individual subsystem elements contribute to a safe system.

Crashworthiness

HSR systems in Asia primarily follow an approach based on *active* safety (i.e., systems that reduce the probability of an accident occurring), whereas in Europe, in addition to active safety, *passive* safety elements are incorporated into the system design (i.e., systems that reduce the consequences of an accident, should it occur). It is generally understood that passive safety systems are not used to compensate for a lack of active safety in the railway network, but are complementary to active safety to account for personal safety when all other measures have failed.

CEM and crashworthiness are design philosophies embraced in passive safety system design. These designs provide a means of dissipating the energy produced during a collision to provide protection to the occupants. CEM systems incorporate collapse zones located in non-occupied areas, typically those close to the extremities of each vehicle, in front of the cab, or adjacent to intercar gangways. The crush zone designs for the Alstom AGV and the Siemens Velaro trainsets are illustrated in Figure 4.

The key safety parameters addressed through crashworthiness include reducing the risk of overriding (e.g., trainsets riding over adjacent units), absorbing collision energy in a controlled manner, maintaining survival space and structural integrity of the occupied areas, limiting the deceleration, and reducing the risk of derailment and limiting the consequences of hitting a track obstruction.

Figure 4. Alstom AGV (Le Corre, 2010) and Siemens Velaro (Schmidt, 2010) front crumple zones

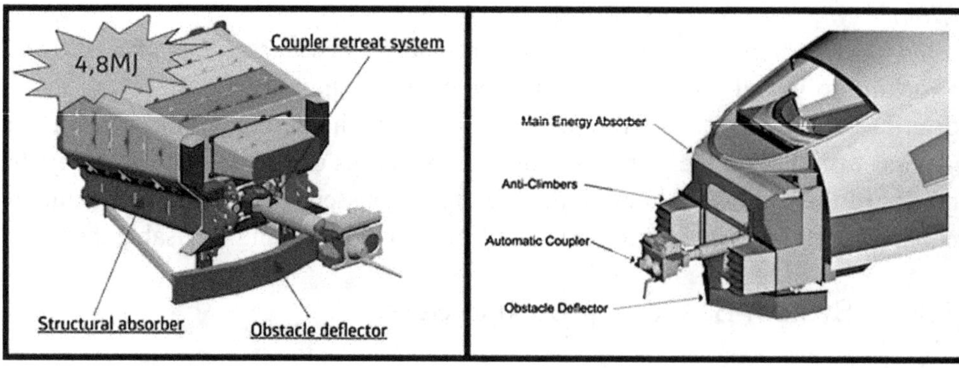

Infrastructure

A common attribute of successful HSR programs is long lengths of straight track connected by curves with generous curve radii because such an alignment maximizes the amount of time a trainset can remain at a desired operating speed, and minimizes wear on the wheels and rail. U.S. HSR projects may be required to operate over existing, and potentially shared, railroad infrastructure to gain access into urban centers, yards, and depots. Although such programs might wish to upgrade existing infrastructure to improve trainset performance and passenger comfort, spatial constraints could dictate that sections of track alignments have small curve radii.

Efforts have been made on establishing the optimal interface between the wheel tread and the rail, which is critical to providing a stable, safe, and comfortable ride. The level of stability required at the highest operating speeds needs to be considered when determining this interface, as does the resistance to derailment at low speeds. Complexity is added when the trainset is operated over dedicated and shared corridors because the rail profiles and maintenance standards may differ.

International HSR trainset manufacturers have stressed the importance of developing lightweight trainset designs to maximize operational efficiency and to minimize maintenance costs associated with the trainset and the infrastructure. The TSI identifies a maximum static axle load of 18.7 tons (17 tonnes). SNCF advised that when it designed high-speed trains in the 1970s, it was critical to control the dynamics of the vertical forces and keep them at the same level as those of conventional trains. In addition, it was essential to minimize unsprung masses (i.e., trainset masses that are not supported by the bogie suspension) because of their potential to generate high forces at high frequencies. As a result, the 18.7 tons (17 tonnes) static axle load specification was set to limit/reduce the effects of those forces, keeping their magnitude comparable to those of a freight wagon running at 62 mph (100 km/h) or a locomotive at 124 mph (200 km/h).

North America's rail operations typically use ground-based axle bearing health monitoring systems (e.g., hot box detectors) to monitor the temperature of a bearing passing above the sensor. An elevated temperature reading results in notification to the operator that an overheated bearing condition may exist. Alstom advised that axle bearing temperature detectors are installed onboard the trainset. Siemens advised that bogie health monitoring systems are mounted on its trainsets. Different alarm levels notify the operator as to when to reduce speed or stop the vehicle for visual inspection.

Traction Power Supply System and OCS

The interactions of high-speed trainsets with traction power and OCS are critical to ensuring stable power supplies to the trainsets and the overall safety and reliability of the HSR system. Trainset-to-OCS interfaces become complicated when trains travel at very high speeds.

China advised that the pantograph-to-OCS interaction was critical to guaranteeing stable current collection. Several technical challenges associated with this interface include the vertical displacement of the pantograph from the OCS, stability of current collection with multiple pantographs in contact with the OCS, effects of aerodynamics and noise, and operations under multi-line conditions and voltages.

China advised that the contact force becomes greater as the speed of the trainset increases. The uplift force is increased proportionally to speed due to aerodynamics. An air deflector near the pantograph was added to help mitigate this effect. During testing to verify proper characteristics of aerodynamics, noise, and resistance, it was discovered that a single-arm pantograph performs better than a double-arm pantograph.

SNCF conducts OCS simulations for studying nominal and perturbed situations and for researching potential optimization techniques. Studying nominal situations enables SNCF to understand better the physics behind HSR operations and the effects that the trainset pantographs have on the OCS and vice versa. SNCF uses this research to optimize the catenary system for potential higher speed operations.

Siemens advised that the pantograph head and the contact strip designs are crucial and project-specific; however, the pantographs are generally designed for speeds of up to 249 mph (400 km/h).

Tunnels

The design of tunnels, especially the cross-sectional area of tunnels, is a key point for integration with the trainset, the operating plan, and the surrounding environment. The tunnel cross-sectional area needs to take into consideration the tunnel length, maximum operating speed through the tunnel, and trainset aerodynamic characteristics. The pressure waves generated by a trainset traveling into and out of a tunnel can adversely affect the trainset carbody in the form of fatigue cracking, and may also result in a sound wave (i.e., sonic boom) being generated at the end of the tunnel. The high cost of tunnels needs to be balanced with the operating plan so that an efficient solution can be realized.

Inspection and Maintenance

A comprehensive and effective inspection and maintenance program is critical to the safe and efficient delivery of HSR service. This program provides assurance that the trainsets are maintained to a high standard of readiness and compliance with regulations. A common theme voiced by operators of HSR systems emphasized that the safety of the system relies upon proper maintenance. Otherwise stated, if one forgets about maintenance, there is no safety. In the U.S., it is typical to establish inspection and maintenance programs using calendar-based milestones. International best practice is to establish maintenance milestones based on mileage, operating hours, and/or actual condition of the component.

In France, condition-based maintenance and remote train monitoring are implemented to assist in advancing HSR maintenance approaches. An example of condition-based maintenance is the overhead monitoring system, which includes laser and optical systems that obtain information on the performance of the pantograph and the behavior of the OCS. Real-time data can be transmitted for monitoring and maintenance planning purposes to improve the overall fleet performance.

Multi-year service contracts can be procured with the trainsets that include preventative maintenance, corrective maintenance, development of a maintenance plan, repair and modification of train components, spare parts management, train assistance along the network, and management of the maintenance depot.

HSR operators invest in trainsets to have them in operation and not in maintenance and depots. The main objective of maintenance is to support operations by providing passenger comfort and safety, train availability and punctuality, and optimized costs. Therefore, a highly optimized maintenance system is needed to realize the true value of HSR. The diagnostic systems for high-speed trains greatly enhance the availability of the fleet by focusing the maintenance effort and ensuring a short maintenance downtime. These systems continuously monitor the train's functions and detect any deviation from normal operation. Internationally, there is a continual improvement in the diagnostic/feedback capability of trainsets to improve reliability.

LEVERAGING INTERNATIONAL BEST PRACTICES AND LESSONS LEARNED

Leveraging Service-Proven Designs for US HSR Programs

NOT reinventing the wheel! The U.S. is in an enviable position with regards to developing HSR in that it can benefit from lessons learned from over 50 years of HSR operation worldwide. To benefit from these lessons, U.S. programs should carefully evaluate their needs, and determine the extent to which service-proven designs and approaches to safe, reliable operation may be applicable. An example of this type of approach can be found in the proceedings of the FRA's Railroad Safety Advisory Committee (RSAC) ETF.

The ETF was convened in 2009 with a stated objective to produce a clear set of criteria and procedures for comparing crashworthiness performance of new trainset designs with compliant FRA Tier I equipment. These criteria and procedures would provide the framework for evaluating alternative rolling stock designs to the requirements identified in FRA Tier I regulations to confirm that an equivalent level of safety, from the perspectives of crashworthiness and occupant protection, is achieved. With this type of approach available, international design philosophies could be evaluated, allowing service-proven designs to be implemented on U.S. programs. The ETF's efforts culminated in the issuance of a document titled *Technical Criteria and Procedures for Evaluating the Crashworthiness and Occupant Protection Performance of Alternatively-Designed Passenger Rail Equipment for Use in Tier I Service (Criteria and Procedures).*

Building on the achievements of the ETF, and in response to the renewed interest in developing a national HSR network, FRA expanded the role of the ETF to develop the framework for a new set of regulations to govern U.S. HSR operations. The new classification for these new regulations was introduced as Tier III.

In October 2010, the ETF began the process of comparing U.S. regulations and best practices against European and Asian approaches to delivering safe HSR. Using methods similar to those used by the ETF in developing the Tier I alternative *Criteria and Procedures*, international rolling stock designs were evaluated to determine if an equivalent level of safety could be provided while preserving the benefits of utilizing service-proven designs and operational approaches. Progress is being made in developing draft regulatory language to address the items below.

U.S. Regulatory Initiatives Support Safe HSR Operation

Structural and Crashworthiness

European trainset designs have been developed based upon the structural requirements prescribed in the Euro Norm EN 12663: *Railway Applications – Structural Requirements of Railway Vehicle Bodies* and upon the performance requirements identified in EN 15227: *Railway Applications – Crashworthiness Requirements for Railway Vehicle Bodies.*

In Europe, passive safety elements are incorporated into the system design. This perspective is evident in the EN 15227 collision scenarios identified below, whereby the crashworthiness design of the trainset is measured against lower-speed impacts, as opposed to a high-speed collision:

- Front-end impact between two identical train units at 22 mph (36 km/h);
- Front-end impact with a different type of railway vehicle, such as an 88-ton (80-tonne) freight car/wagon at 22 mph (36 km/h);
- Train unit front-end impact with a large road vehicle, such as a 16.5-ton (15-tonne) truck/deformable obstacle at an at-grade crossing at a maximum of 68 mph (110 km/h);
- Train unit impact into a low obstacle, such as a car, animal, or rubbish, or an at-grade crossing to achieve obstacle deflector requirements.

The Tier I *Criteria and Procedures* incorporated trainset structural and CEM/crashworthiness attributes based upon European best practices. Using predefined collision scenarios, it identifies criteria that, if adopted, will provide an equivalent level of crashworthiness to current Tier I requirements, while allowing the use of service-proven, lightweight and efficient rolling stock designed to international standards.

The current development of Tier III structural and crashworthiness standards also draws upon international experiences, while providing an equivalent level of safety as current regulatory requirements, when these trainsets are operated in Tier I territories. Areas being addressed include occupied volume integrity (which is intended to demonstrate resistance to loss of occupied volume via criteria for quasi-static compression loads and for a dynamic collision scenario), anti-climbing performance (from both colliding and connected equipment perspectives, evaluated during the dynamic collision scenario), inhibition against fluid entry, integrity of the cars (e.g., collision and corner post requirements for the cab and non-cab ends), integrity of the roof and side structures, and truck attachment strength.

Glazing: Forward-Facing

The safety of the train crew is of paramount importance. One element of train design directly related to the safety of the crew is the design of the forward-facing cab windscreen.

There are numerous design attributes associated with high-speed trainset cab windscreens including: protection of the train crew from external threats (e.g., airborne debris, birds, and animals); improvements on aerodynamics of the train; reduction in the generation of exterior noise; and allowance for acceptable visibility to see wayside-mounted signal indications. Contemporary high-speed trainsets incorporate a monolithic shaped windscreen (refer to Figure 5) into the design of the end of the trainset.

As shown in Figure 6, the optical thickness of glazing increases with the reduction of the angle of installation. As the optical thickness increases, optical clarity decreases. Therefore, manufacturers of high-speed train windscreens look at ways to construct these windscreens by using strengthened glass layered with polymer interlayers to achieve the appropriate balance of optical clarity and impact resistance.

The ETF is evaluating international best practices associated with the design and testing of forward-facing cab windscreens to facilitate development of Tier III

Figure 5. CRH380A trainset with monolithic shaped windscreen (Li, 2010)

Figure 6. Illustration of optical thickness related to glazing thickness and installation angle

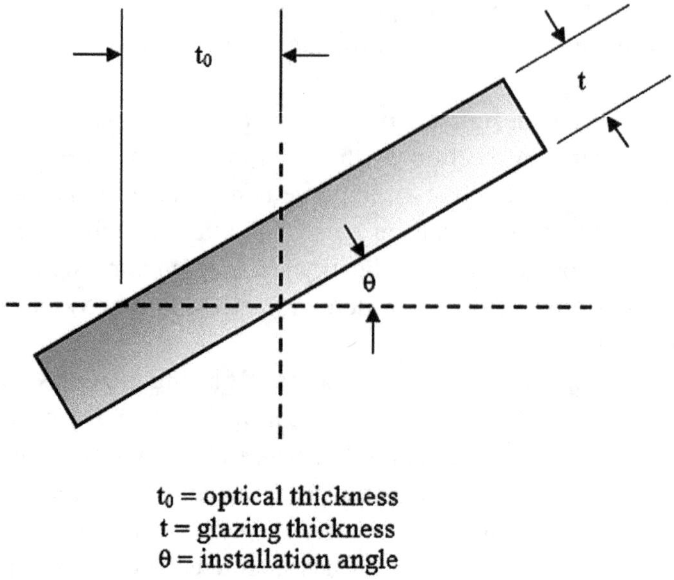

safety regulations that will result in acceptable levels of train crew safety and windscreen performance.

Glazing: Side-Facing

The U.S. approach to emergency windows typically incorporates the use of an elastomeric gasket that can be removed from the perimeter of the window to allow passengers to dislodge the window without the use of tools (refer to Figure 7). The international approach is to have passengers use a hammer-type device to break a specially designed emergency window (refer to Figure 8). Once the window is broken by the passenger, the entire unit can be pushed out of the trainset side wall.

HSR trainsets will encounter significant pressure pulses during operations that could impact side-facing windows. Such pulses are typically the most severe when two trainsets pass each other on closely spaced track centers or when a trainset enters and leaves a tunnel. To prevent a window from being dislodged, European manufacturers design side-facing windows to withstand the pressure pulses anticipated during operation. These designs typically preclude the use of an elastomeric gasketed window and frame assembly in favor of an assembly that is mechanically fastened to the carbody.

Figure 7. Typical elastomeric gasketed side-facing emergency window (FRA, 2010)

Figure 8. Velaro E emergency side glazing with breakout hammer (Banko, 2010)

FRA recognizes the merits of the international design philosophy for breakable side-facing emergency windows, and the associated records of safe operation. The installation of breakable side-facing emergency windows in lieu of the traditional elastomeric gasket design is likely to be permitted, provided that an equivalent level of safety is demonstrated.

Braking

High-speed trainsets incorporate the use of electric and mechanical friction-based braking systems to stop safely and reliably within the required distances. Equipment manufacturers have refined their systems to accommodate the mechanical and thermal stresses associated with stopping trainsets from very high speeds. HSR operators have evaluated the performance aspects of the braking systems to determine the most reliable and cost-effective approaches. Trainsets typically employ electric braking at high speeds, using only the friction brakes during low speeds (e.g., 44 mph (70 km/h) or less) or during an emergency.

Key attributes of service-proven high-speed trainset brake design and operation have been evaluated by the ETF. One attribute allows for driver intervention of the emergency brake application; international best practice is to allow for the train operator to intervene in the application of the emergency brake under defined conditions. For instance, should a fire occur onboard a trainset, the operator is in the best position to determine the safest location for a train to stop (e.g., potentially outside a tunnel). Another attribute is thermal capacity, where international best practice is to design the brake system to be able to safely stop the trainset within pre-defined distances, using only friction-based braking, as opposed to taking benefit of the electric braking capabilities for the trainset. A third attribute relates to the provision of a wheel-slide control system/locked axle detection system, whereby contemporary high-speed trainsets are provided with redundant systems to alert the operator in the event of a non-rotating wheelset. Lastly, there is brake system health monitoring, where these trainsets employ a diagnostic system to monitor the performance of the system and to alert the operator of issues that could result in degraded performance.

The ETF has been assessing service-proven braking systems and associated operational practices and developing draft regulations that leverage these best practices for the U.S. application. This also includes review of the current regulatory language, whereby some of the existing CFR criteria (e.g., main reservoirs) may be retained. To leverage the use of service-proven high-speed rail equipment, consideration is also given towards permitting the railroad to present preferred approaches, for FRA consideration, while justifying that the safety of the system is maintained.

Interior Fittings and Surfaces

U.S. and international design practices have included the requirement for interior equipment to be installed in such a manner so that the equipment mountings can withstand the forces of deceleration resulting from a derailment or collision. The ETF has been evaluating European best practices relating to the prediction of the forces that would result from EN 15227 type collision scenarios, and the extent to which the CEM system will reduce these forces. It is being considered that the Tier III criteria will accommodate the implementation of international best practices relative to the strength of interior fittings, provided that the standards are implemented in accordance with the railroad's FRA approved system safety plan.

One example of a difference between current regulations and what is being assessed for Tier III is the requirement for the luggage racks. U.S. regulations for Tier II trainsets require luggage racks to be enclosed. This is contrary to international best practice where it is recommended that luggage racks are not enclosed, but are designed to incorporate semi-transparent materials that allow train crew to readily detect packages left behind. International operators cite concerns where articles can be left behind that may pose a threat to passenger safety. The ETF is proposing draft regulations that will allow this international best practice to be implemented.

Safety Appliances

U.S. regulations cite numerous requirements relating to safety appliances installed on trains. These safety appliances include handrails, hand holds, ladders, and steps that are used by the train crew to gain access to the train, and to perform necessary activities. The safety appliance regulations have been developed based on train designs that have operated in North America, and are prescriptive in nature in that minimum and maximum dimensions are identified for placement of these safety appliances.

The ETF has been evaluating the existing regulations against the designs of service-proven high-speed trainsets and has identified instances where the currently prescribed dimensions are not compatible with the structural design of these contemporary trainsets. The ETF is proposing draft regulation for safety appliances installed on these trainsets that will provide an equivalent level of safety for train crew while accommodating international best practices associated with trainset structural design.

Inspection, Testing, and Maintenance Requirements

High-speed trainsets are designed to incorporate systems that monitor the health of key systems and the overall performance of the train. These monitoring and diagnostic systems can be leveraged to improve efficiencies associated with the inspection, testing, and maintenance of the equipment. For example, condition monitoring of critical components can provide alerts to maintenance personnel that action is needed to be taken prior to a component failure occurring. As the component is degrading, replacement parts and the required maintenance resources can be staged to rapidly return the train to a state of operational readiness. Historical trends developed from the data collected from the monitoring systems can be referenced to inform the scheduling of inspection and testing activities.

As with the other elements identified above, the ETF is evaluating opportunities for leveraging international best practices associated with the inspection, testing, and maintenance of high-speed equipment.

Emergency Systems

Emergency systems have also been evaluated by the ETF to determine applicability and appropriateness for Tier III trainsets. In addition to the emergency windows for access and egress, requirements for emergency lighting have been assessed, which includes attachment strength and operational performance of backup systems after a collision or derailment.

One emergency system that may need to be addressed, however, is emergency door opening. U.S. regulations dictate that all coaches be equipped with manual override devices accessible from the interior and the exterior of the trainset. These devices must be installed adjacent to the doors that they control. U.S. design and performance requirements for these devices include a prohibition against incorporating an interlock signal for actuation of door override devices. International practice typically includes a speed interlock (either 3 mph (5 km/h) or 6 mph (10 km/h)), whereby the door may be opened under emergency conditions. Refer to Figure 9 for the emergency door opening device installed on the Siemens Velaro E trainset.

The use of manual door override devices were discussed with the Spanish HSR operator, Renfe. Renfe advised of concerns relating to opening exterior doors (via the manual door override) while the train is in operation, recounting a passenger who committed suicide by opening the door of a trainset while the train was moving at a high speed. Renfe prefers the trainsets to apply a brake application automatically when a passenger opens the door via the emergency door release. Renfe also believes that such an action should be difficult (to prevent unwarranted door openings) for passengers, and that the doors should not be allowed to open while the train is moving.

While the manual override device and its associated speed interlock is currently not a topic of discussion for the ETF, the feasibility of employing such an interlock would be appropriate for evaluation at the project level.

Figure 9. Velaro E emergency door opening device (Banko, 2010)

CHALLENGES MOVING FORWARD

While there has been much progress made and many of the challenges are being resolved or have been resolved, there is still much work to be done. In particular, there are three key issues that will need to be addressed in the near future as HSR progresses in the U.S. – trainset weight, vehicle-track interaction (VTI), and exterior noise.

Trainset Weight

Static Axle Load Requirements

One of the key attributes of a safe and efficient high-speed operation is the use of lightweight equipment that limits the forces exerted on the rail. This attribute translates to the static axle load of the trainset. Benefits of a reduced axle load include, but are not limited to, improvement in energy efficiency, reduction in rolling noise, reduction of wear and tear on the infrastructure and rolling stock, and associated reduction in maintenance costs.

The TSI sets forth the criteria for measuring static axle load, which include normal load conditions inclusive of:

- Passengers in seating areas, including those seats in bistros/dining cars;
- Compartments containing luggage, freight, etc., whereby the maximum payload expected to be encountered in revenue service is to be factored;
- Train crew;
- All materials required for operation, such as lubricants, coolants, catering equipment, fresh and waste water, etc.; and
- Two-thirds of consumable items, such as sand, food, etc.

For equipment that operates at a speed greater than 155 mph (250 km/h), the TSI identifies a nominal static axle load of 18.7 tons (17 tonnes). Table 2 offers a comparison of the axles loads of several high-speed trainsets. It is noted that the requirements identified above must be met regardless of whether the trainset utilizes conventional or articulated bogie architecture[6]. As such, it is not uncommon to see trainsets with an articulated configuration (e.g., Alstom's TGV and AGV and Hyundai Rotem's KTX-II) already at the TSI limits.

Table 2. Comparison of static axle loads of various high-speed trainsets

	TGV	AGV	Velaro	N700	CRH380A	KTX-II	Acela
		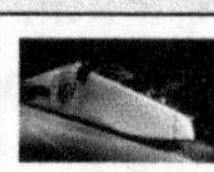	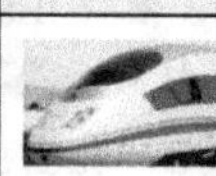	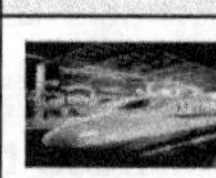	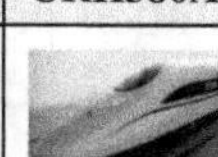	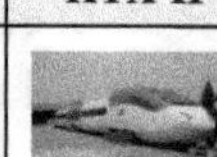	
Axle Load, tons (tonnes)	18.7 (17)	18.7 (17)	14.3 (13)	12.3 (11.2)	16.5 (15)	18.7 (17)	23 (21)

Projections in Passenger Weights

The representative weights of the passengers, luggage, and crew members are a factor to ensuring that the trainset is designed to be structurally sound. These weights are also to be accounted for in the determination of the trainset's static axle load as described above. The representative weight of passengers, inclusive of luggage, used in trainset design per TSI and EN 15663 is 176 lb_m (80 kg). In Japan, a 121 lb_m (55 kg) value is used.

In the U.S., 2013 Center for Disease Control and Prevention (CDC) data identified the average weights of males and females as 195.5 lb_m (88.7 kg) and 166.2 lb_m (75.4 kg), respectively. These values already exceed the TSI criteria and recent trending has shown that they are expected to rise; as such, it is appropriate that the trainset be designed to accommodate the weights expected to be seen over the course of the equipment's life (e.g., 30 or 40 years). Assuming an increase of 0.65 lb_m/year (NCSF, n.d.) for the next 30 years, anticipated weights for males and females, respectively, will be 215.0 lb_m (97.5 kg) and 185.7 lb_m (84.2 kg) (refer to Figure 10). Additional weight will also need to be factored in for luggage.

Moving Forward

While several of the high-speed trainsets are already at the 18.7 ton (17 tonne) axle load limit, it is critical that the equipment being designed meets the anticipated operational needs over its lifetime. This should include, at a minimum, the appropriate projected weights expected to be hauled not only at inception of service, but also by the end of its expected life (typically 30 or 40 years). In addition, for the U.S., accommodations for additional weight may also be needed as the new Tier III criteria are implemented, especially with regards to implementing the proposed crashworthiness and structural requirements.

Figure 10. U.S. weight projections to 2043

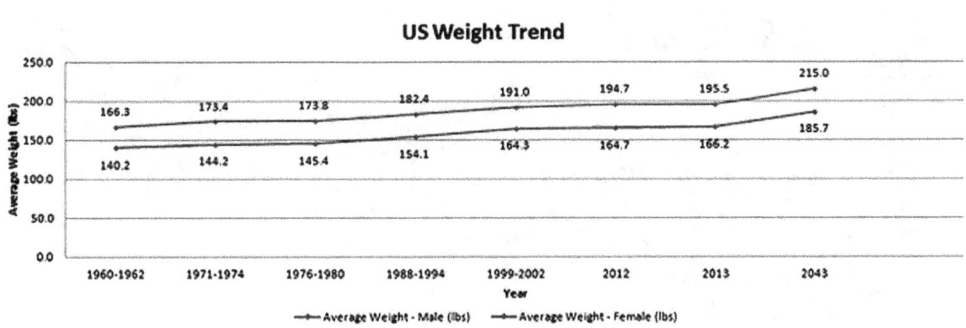

Vehicle-Track Interaction

HSR Bogie and Suspension Designs

Many high-speed trainset designs employ the use of lightweight, bolsterless bogies. The primary suspension is typically a spring (e.g., coil, leaf), whereas the secondary suspension is typically pneumatic (e.g., air bags). Figure 11 shows an example of the SF 500 bogie used on Siemens' Velaro trainsets.

High-speed trainsets are designed with one of three bogie architectures (refer to Figure 12). One of these is the conventional architecture, whereby each car in the trainset is equipped with two bogies - one on each end; examples of high-speed trainsets that employ the conventional bogie configuration include the Velaro, Shinkansen, ETR 450, and the Pendolino. The second is the semi-articulated architecture, whereby the bogies are not shared between the leading/trailing cars and their adjacent cars, but are shared in the remaining intermediate cars; examples of high-speed trainsets that employ the semi-articulated configuration include the TGV and the Talgo 350. The last is the fully articulated architecture, whereby each bogie is shared by two cars; an example of a fully articulated trainset is the AGV.

Several trainsets are also designed with tilt mechanisms, which allow the equipment to negotiate curves at a higher speed, while reducing the centrifugal forces felt by the occupants. The two types of tilting employed are active tilt (tilting via control by an onboard computer) and passive tilt (tilting caused by the vehicle's inertial forces). Benefits of having tilting equipment can be realized on existing

Figure 11. Siemens SF 500 high-speed bogie (Schroeder-Bodenstein, 2010)

Figure 12. Conventional vs. articulated bogie architectures (Avril, 2010)

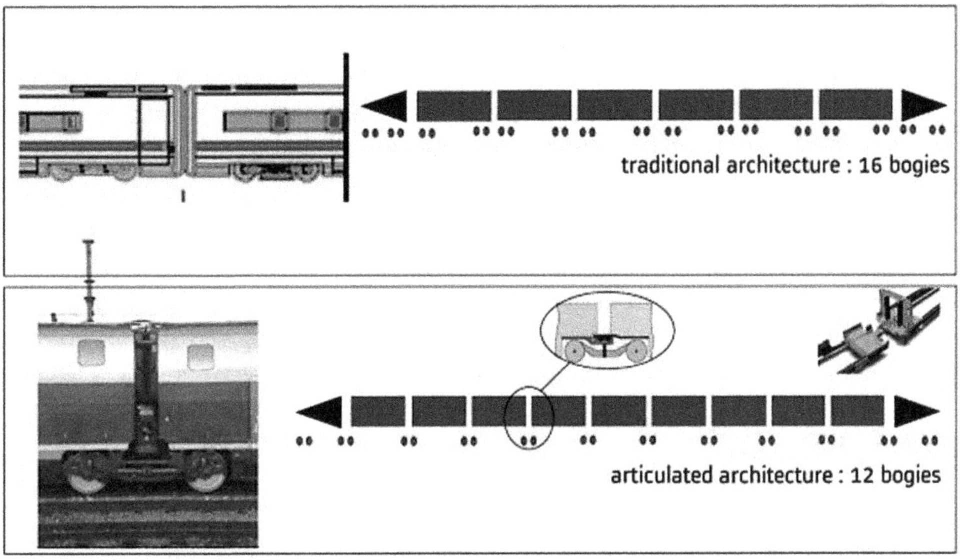

infrastructure with tight curve radii. Tilting designs, however, are only proven up to 168 mph (270 km/h) and typically require complex assemblies (particularly those with active mechanisms).

FRA Track Classes and Vehicle-Track Interaction Safety Standards

In March 2013, FRA published the document titled *49 CFR Parts 213 and 238 Vehicle/Track Interaction Safety Standards; High-Speed and High Cant Deficiency Operations; Final Rule*. The intent is to "[amend] the Track Safety Standards and Passenger Equipment Safety Standards to promote the safe interaction of rail vehicles with the track over which they operate under a variety of conditions at speeds up to 220 mph [(354 km/h)]. The final rule revises standards for track geometry and safety limits for vehicle response to track conditions, enhances vehicle/track qualification procedures, and adds flexibility for permitting high cant deficiency train operations through curves at conventional speeds" (FRA, 2013).

Cant deficiency operations involve traversing through curves at speeds greater than the balance speed. In Europe, cant deficiency is typically limited to 5 inches (127 mm) on high-speed lines. In France, for comfort purposes, cant deficiency is typically limited to 1.6 to 2 inches (40 mm to 50 mm). In the Infrastructure TSI, cant deficiency is limited to 7.1 inches (180 mm) for speeds up to 99 mph (160 km/h); for speeds above 186 mph (300 km/h), cant deficiency is limited to 3.1 inches (80 mm).

As part of the ETF, a VTI subtask group was formed in 2011 to evaluate how key track parameters affect performance and safety for service-proven high-speed equipment. Comparisons were highlighted between the U.S. and international track geometry standards and efforts were made to identify potential issues arising from operation of Tier III equipment over Tier I track. Industry had expressed concerns that service-proven HSR trainsets may not be able to negotiate track that is maintained to standards categorized as FRA track classes 1 through 3 (refer to Table 1). Progress was made, and equipment manufacturers performed simulations to confirm trainset performance over the various FRA track classes. However, based on the proprietary nature of HSR trainset design, and the inability for industry and the FRA to reach consensus on a path forward, the VTI subtask group was disbanded, with the agreement that VTI issues would be evaluated on a project-by-project basis.

Moving Forward

Due to differences in operating environments and each manufacturer's unique bogie and suspension designs, it is critical that VTI be evaluated at the project level where manufacturers can familiarize themselves with project system requirements; fine-tune their designs to meet such criteria, and advise of any issues with meeting FRA's safety standards, while maintaining the attributes of a service-proven design.

Exterior Noise

Comparison of Noise Limits

Exterior pass-by noise remains a challenge for HSR in the U.S., especially when it comes to operating at the higher speed regimes. The challenge lies in maintaining these emissions to current U.S. regulatory levels, which have historically been defined around the basis of capturing noise from a diesel locomotive engine and are independent of speed. Based on research conducted to-date, it is anticipated that high-speed trainsets will exceed the current EPA/FRA guidelines, as codified in 40 CFR §201 and 49 CFR §210, for exterior noise generation when speeds of 199 mph (320 km/h) and higher are attained.

Per 40 CFR §201.1, a high-speed trainset is defined as a "locomotive." As such, the requirements of 40 CFR §201.12 apply, whereby the maximum noise level must not exceed 90 dB(A) at any time during a pass-by event, when measured 100 feet (30 m) from the centerline of track and 4 feet (1.2 m) above the ground.[7]

The TSI offers an alternative approach towards determining maximum sound levels during pass-by events. The TSI utilizes the "mean" of the maximum noise levels during a single event, when measured 82 ft (25 m) from the centerline of track and 11 ft (3.5 m) above top of rail (measuring at this height accounts for the emissions from the pantograph, which represents one of the primary sources of exterior noise due to aerodynamics).[8] The TSI values for pass-by noise are speed-dependent, with higher limits allowed for higher speeds, as shown in Table 3.

With the difference in receptor locations identified by the CFR and the TSI, it is determined that the CFR equivalent to the 93 dB (A) identified in Table 3 for 199 mph (320 km/h) is 91.4 dB (A), which exceeds the EPA limit.[9] To reasonably estimate the sound level at 220 mph (354 km/h), an acceptable practice is to add 1 dB(A) for every 6 mph (10 km/h) increase in velocity above 199 mph (320 km/h); as such, the CFR limit at 220 mph (354 km/h) will be approximately 94.4 dB(A) when considering the TSI limits.

Unlike the U.S. and European metrics, in Japan, the noise requirements factor in wayside sound mitigation measures (e.g., noise barriers) and the location of actual ground (e.g., street level). For the Shinkansen, the Ministry of the Environment calls for 75 dB (A) or less depending on the environment through which the trainset is operating (e.g., residential or commercial/industrial). The limits are to be satisfied when measured 82 feet (25 m) from the centerline of the nearest track and 4 feet (1.2 m) above the ground. Implementing wayside measures to counter noise is one example of a system-based approach towards mitigating the impact to the surrounding environment.

Exterior Noise Sources

Exterior noise from a high-speed trainset can be categorized three-fold. The first is aerodynamics, which involves noises from the upper portions (e.g., pantograph), the lower portions (e.g., bogies), the intercar locations, and the nose of the trainset; this category also includes noises generated by non-flush exterior components (e.g., safety appliances) and crevices (e.g., bogie areas). As shown in Figure 13, aerodynamic noise is the predominant factor of pass-by noise for speeds above 186 mph (300 km/h)[10]. It is noted, however, that between 186 mph (300 km/h) and 218 mph (350 km/h), rolling noise would be dominant if the wheels are worn and the wheel and rail interaction is poor. Rolling noise involves contact noise due to vehicle-track interaction; this source also depends on the trainset weight, the

Table 3. Limits for pass-by noise per TSI section 4.2.6.5.4

	155 mph (250 km/h)	186 mph (300 km/h)	199 mph (320 km/h)
$L_{pAeq,Tp}$, dB(A)	88	92	93

Figure 13. Sound pressure levels from various sources (Cuny, 2010)

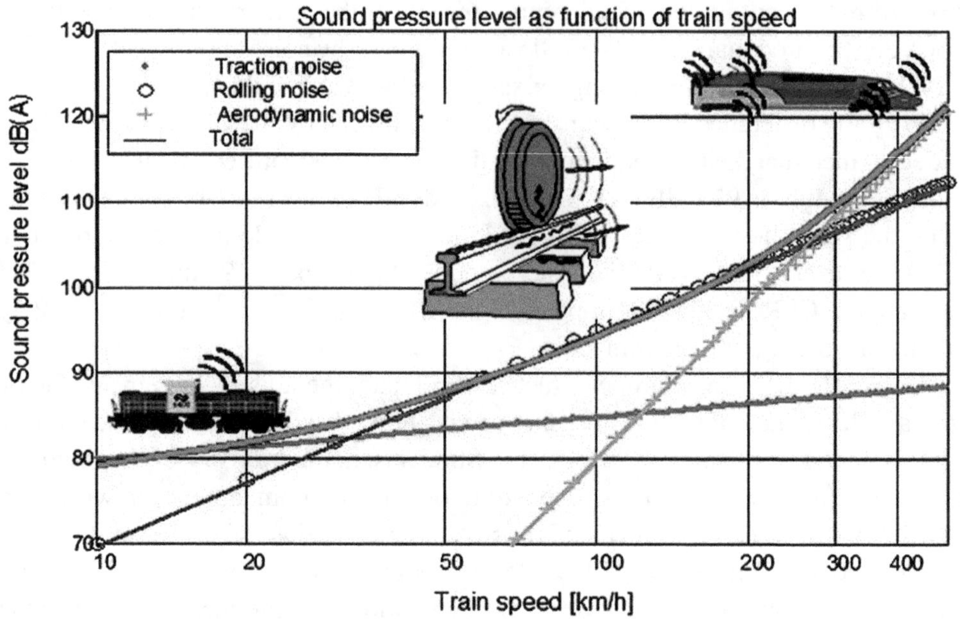

method of track fixation (there is potential for a +3 dB (A) delta for trains operating over slab track) and the surface conditions of the wheel and rail. The last source is equipment noise, which involves noises and vibrations generated, for example, by traction equipment, heating, ventilation, and air conditioning (HVAC) equipment, and pantograph-to-OCS contact.

In addition to train-borne noises, there is noise generated by the vibration of the supporting structure during a pass-by event.

Onboard and Wayside Mitigation Measures

Manufacturers are constantly evolving their trainset designs to provide the most optimal solutions for HSR. Measures to decrease aerodynamic noise emissions onboard the trainset include the use of streamlined front ends, intercar shrouds, flat underfloor panels, bogie covers, pantograph covers and noise insulation barriers, low-noise and single-arm pantographs, and lower carbody sound absorbing panels (see Figure 14). To mitigate rolling noise, trainsets are being developed with lower weights (i.e., reduced axle loads).

With the pantograph being the most conspicuous source of aerodynamic noise, the Japanese in particular have been making strides towards improving the current collection system so as to reduce the level of emissions during high-speed opera-

Figure 14. Examples of onboard noise mitigation measures (Kurita, n.d.)

tions. Designs on mitigating noise have evolved throughout the years and the merits of this research can be seen installed on the latest Shinkansen trainsets. The modifications of using noise insulation barriers and the modified pantographs have resulted in a noise reduction of more than 2 dB(A) in pantograph peak noise level, when compared to the design of the earlier E2 Series Shinkansen (Hara, 2009).

On the wayside, HSR corridors are developed with wind walls/sound barriers to alleviate the impact of pass-by noise on the surrounding areas (see Figure 15). The intent of using these barriers is to permit absorption (into noise-absorbing panels) and reflection (thereby causing destructive interference) to limit the noise encountered at the receptor.

Figure 15. Wind walls/sound barriers on the ICE line (Koschnitzke, 2010)

Moving Forward

With aerodynamics playing a major role in noise levels at these higher speeds, it is critical to revisit the existing CFR criteria so as to ensure that operation at these speeds can be realized. One consideration could be the development of a new definition for "high-speed trainset" in the CFR, whereby new speed-dependent limits are identified. This would be consistent with the TSI approach.

OPPORTUNITIES

The quest to forward HSR operations within the U.S. continues. Looking back at steam locomotives purported to have reached speeds of 112.5 mph (181 km/h), to Amtrak Acela Express trainsets capable of operating up to 150 mph (241 km/h), and now forward to high-speed electric trainsets capable of operating up to 220 mph (354 km/h) for projects like the California High-Speed Rail Project, HSR continues to advance in the U.S.

As proven through European and Asian high-speed operations, HSR still remains a viable and popular option for travelers looking to cover the distance typically traversed by short-haul airplanes. Through various Greenfield HSR projects around the country and the efforts to upgrade existing lines (and the already high-speed NEC), the opportunity for an interoperable U.S. high-speed network remains. In addition, as technology continues to advance, it is expected that these trainsets will become even more sustainable, not only from an environmental or efficiency perspective, but also from the business case of supporting a domestic HSR industry.

Interoperability

Since its inception with the Tokaido Shinkansen in 1964, HSR has been offering travelers a safe, energy-efficient, and viable alternative to short-haul air travel. The Japanese high-speed system has since grown with the openings of the Sanyo, Tohoku, Joetsu, Hokuriku, and Kyushu Shinkansens, with the additional Hokkaido Shinkansen currently under construction. Interoperability with the former narrow-gauge lines (Akita and Yamagata Shinkansens) is realized through the use of E3 and E6 Series trainsets, which also serve the main Tohoku Shinkansen.

Across the East China Sea, China has been one of the fastest developers of HSR. Owner of the largest and longest HSR network, China has also been one of the first countries to operate its high-speed trainsets at speeds up to 218 mph (350 km/h). Learning from its various trainset procurements, China has also been able to design and optimize their equipment utilizing best practices from around the world. In China, the newly built track and the existing track are interoperable.

In Europe, HSR has steadily grown since the 1980s into a multi-country network, whereby efforts are in place to provide standardization of HSR to permit interoperability among the different countries. The fruits of these efforts can be seen in the adoption of the TSI, which offers a comprehensive view of the HSR system, inclusive of the rolling stock technology.

In the U.S., the commitment towards rail improvement can be witnessed via the ARRA program. Ongoing projects, such as the California High-Speed Rail Project, All Aboard Florida, and Xpress West coupled with FRA's efforts behind updating and developing regulatory text for Tier I and Tier III operations, respectively, continue to show the support for HSR, and provide an opportunity for the U.S. to one day have an interoperable high-speed network.

Sustainability

One of the benefits of HSR is the promise of sustainability from both technological and business perspectives. Rolling stock manufacturers are constantly making sustainability one of their core areas of improvement, while rail agencies are looking to provide efficient and cost-effective transportation solutions.

Technological sustainability of high-speed equipment is realized through the reduction of energy consumption and CO_2 emissions, and through optimized energy efficiency via improved trainset designs. For example, in fuel equivalent terms, the Alstom AGV consumes only 0.4 L of fuel per 100 passenger-km, which is three times less than a bus (1.2 L), eight times less than a car (2 to 3 L), and 15 times less than an airplane (5 to 7 L) (Alstom, n.d.). In addition, this trainset emits less than 10 g of CO_2 per passenger-km, as compared to 81.3 g/passenger-km for personal vehicles and 243.8 g/passenger-km for airplanes (Alstom, 2009). Siemens' Velaro trainset consumes 0.33 L of fuel per 100 passenger-km.

With respect to improvements in trainset design and onboard technologies, efforts are made to reduce the axle load of the equipment (e.g., use of aluminum extrusions as opposed to stainless steel), improve aerodynamics (so as to reduce drag and noise emissions), and reduce net energy consumption (through the use of regenerative brakes and automatic train operation). Furthermore, manufacturers are developing modular technologies (e.g., Bombardier's ECO4 – ecology, economy, efficiency, and energy) that would further enhance energy savings and improve overall efficiency and performance.

From the business perspective, with the support of Buy America and subsequent proper transfer of technology, a new domestic industry for high-speed trainset manufacturing and assembly can be born. As the U.S. continues upgrading its rail infrastructure to support these higher speeds, this new market can flourish in the midst of providing alternatively designed Tier I equipment and new Tier III trainsets. With continued support for rail improvement, this market can help create and sustain jobs (down to the trainset component and subsystem levels) with the assurance that safe, efficient, and proven equipment is provided. Furthermore, this new U.S. industry could be positioned to support the manufacturing and assembly of equipment for export.

CONCLUSION

There is still much that the U.S. can learn from its international colleagues and their decades of HSR design and operating experiences. To date, it has been, and continues to be, a joint, international effort in establishing the benchmark requirements for domestic HSR. It is important that the U.S. HSR industry continues to foster dialogue with its international colleagues as to continue to address the challenges and realize the opportunities of HSR development.

REFERENCES

Alstom. (2009). *AGV: Full-Speed Ahead Into the 21st Century.*

Alstom. (n. d.). *The AGV, a New and Revolutionary Very High Speed Train.*

Avril, E. (2010). *Articulated/Conventional Architecture. In 2010.* La Rochelle, France: William Barclay Parsons Fellowship.

Banko, F., & Xue, J. (2012). *Pioneering the Application of High Speed Rail Express Trainsets in the United States (50).* New York, NY: Parsons Brinckerhoff Group, Inc.

Cuny, N. America Referential for High Speed. In 2010 William Barclay Parsons Fellowship. La Rochelle, France.

Federal Railroad Administration. (2009). *High-Speed Passenger Rail Safety Strategy*, 1-2.

Federal Railroad Administration. (2009). *Vision for High-Speed Rail in America, 1*, 6.

Federal Railroad Administration. (2010). *RSAC Engineering Task Force.* Cambridge, MA.

Federal Railroad Administration. (2013). Summary. In 49 CFR Parts 213 and 238 Vehicle/Track Interaction Safety Standards; High-Speed and High Cant Deficiency Operations; Final Rule, 16052.

Hara, M., Kurita, T., Mizushima, F., Satoh, H., Shikama, T., Wakabayashi, Y., & Yamada, H. (2009). Reduction of Pantograph Noise of High-Speed Trains. *Journal of Mechanical Systems for Transportation and Logistics*, *3*(1), 73.

Knowlton, B. (2009). Obama Seeks High-Speed Rail System Across U.S. *New York Times*. Retrieved from http://www.nytimes.com/2009/04/17/us/politics/17train.html?_r=0

Koschnitzke, A., Schüler, J., & Waldeck, H. (2010). *Acoustics. In 2010*. Krefeld-Uerdingen, Germany: William Barclay Parsons Fellowship.

Kurita, T. & Mizushima, F. (n. d.). Environmental Measures along Shinkansen Lines with FASTECH360 High-Speed Test Trains. *JR East Technical Review, 16*, 48, 49, 51, 53.

Lacôte, F. (2010). *Very High Speed Train and Safety. In 2010*. Paris, France: William Barclay Parsons Fellowship.

Le Corre, D. (2010). *Passive Safety in Alstom High Speed Trains. In 2010*. Reichshoffen, France: William Barclay Parsons Fellowship.

Li, Y. (2010). *Information about High Speed EMU. In 2010*. Qingdao, China: William Barclay Parsons Fellowship.

National Council on Strength and Fitness. (n. d.). Obesity Awareness Month. Retrieved from http://www.ncsf.org/enew/articles/articles-obesityawarenessmonth.aspx

Roberts, S. (2013). All Aboard. In *Grand Central – How a Train Station Transformed America (114)*. New York, NY: Hachette Book Group.

Schmidt, G., & Waldeck, H. (2010). *Crashworthiness. In 2010*. Krefeld-Uerdingen, Germany: William Barclay Parsons Fellowship.

Schroeder-Bodenstein, K. (2010). *Running Dynamics and Gauging. In 2010*. Krefeld-Uerdingen, Germany: William Barclay Parsons Fellowship.

ADDITIONAL READING

Code of Federal Regulations. (n. d.). Retrieved from http://www.ecfr.gov/cgi-bin/ECFR?page=browse

High-Speed Rolling Stock Technical Specifications for Interoperability. (2008). Retrieved from http://www.era.europa.eu/Document-Register/Pages/HS-RST-TSI.aspx

Technical Regulatory Standards on Japanese Railways. (2012). Retrieved from http://www.mlit.go.jp/english/2006/h_railway_bureau/Laws_concerning/14.pdf

Technical Specifications for Interoperability. (n. d.). Retrieved from http://www.era.europa.eu/Core-Activities/Interoperability/Pages/TechnicalSpecifications.aspx

KEY TERMS AND DEFINITIONS

Active Safety: Measures that are part of the railway system that reduce the probability of an accident occurring.

Balance Speed: The speed where the lateral component of the centrifugal force is compensated by the corresponding component of the gravitational force.

Bogie: Vehicle component comprised of the frame and wheelsets. Bogies may be powered (motor bogies) or unpowered (trailer bogies).

Cant Deficiency: The difference between the applied cant on the track and the equilibrium cant for the vehicle at the particular stated speed.

Concentrated Power: Trainset/consist configuration whereby the traction equipment is located at the leading and/or trailing ends.

Distributed Power: Trainset/consist configuration whereby the traction equipment is located throughout the coaches.

Electric Brakes: Includes both dynamic brakes and, if installed, eddy current brakes. Dynamic brakes utilize the kinetic energy of the train to generate current at the traction motors; after providing the necessary braking effort, the remaining energy is either returned to the OCS (regenerative braking) or dissipated via onboard resistor grids (rheostatic braking). Eddy current brakes utilize electromagnetic induction imparted on the rail or an axle disc to generate the forces needed to decelerate the train.

Passive Safety: Measures onboard the trainset that mitigate the effects of an accident, should one occur.

Pantograph: Current collection device installed onboard the roof of electric trainsets.

Positive Train Control (PTC) System: As defined by the FRA to reliably and functionally prevent: (i) train-to-train collisions – including collisions between trains operating over rail-to-rail at-grade crossings; (ii) overspeed derailments, including derailments related to railroad civil engineering speed restrictions, slow orders, and excessive speeds over switches and through turnouts; (iii) incursions into established work zone limits without first receiving appropriate authority and verification from the dispatcher or roadway worker in charge; and (iv) the movement of a train through a main line switch in the improper position.

Push-Pull: Configuration whereby a locomotive is either placed at the end of the consist and the driving cab is at the opposite end (push), or at the front of the consist and serves as the driving cab (pull).

Receptor: Microphone situated at a set location from the track and used to measure the pass-by noise.

ENDNOTES

1 A fourth tier, Tier IV, is being considered to support a fully dedicated HSR corridor (e.g., the corridor being planned for the Texas Central Railway).

2 Amtrak has submitted a waiver request to the FRA seeking approval to operate the Acela trainsets up to 160 mph (258 km/h) in revenue service.

3 Refer to FRA Buy America Frequently Asked Questions at: http://www.fra.dot.gov/Page/P0391.

4 Safety assurance is the reduction of hazard risks to an acceptable level of safety.

5 A safety management system is the planning, organizing, directing, and controlling of those activities necessary to achieving an organization's loss-prevention and loss-control objectives.

6 Articulated architecture is a trainset configuration whereby a single bogie is shared by two adjacent coaches. In an articulated configuration, the trainset has less bogies to distribute its mass.

7 Refer to 40 CFR §201.24 for information pertaining to additional test conditions.

8 Refer to TSI Section 4.2.6.5.4 and EN ISO 3095 for information pertaining to additional test conditions.

9 This assumes that the pass-by noise level, as defined as $L_{pAeq,Tp}$ by the TSI, is equivalent to the maximum noise level, as defined as L_{max} by the CFR.

10 Note: The measurements shown were made at a distance closer than the 82 ft (25 m) mandated by the TSI.

Chapter 6
Safety of Domestic High Speed Passenger Rail Operations:
Safety Is Good Business

Stephen C. Laffey
DuPage Railroad Safety Council, USA

ABSTRACT

This chapter examines safety related to high and higher speed inter-city passenger rail operations. Generally, most risk related to rail travel is associated with collisions between on-track equipment and vehicles and pedestrians at level crossings. A larger amount of fatalities, but not necessarily systemic risk, is due to trespassing on the rail right-of-way. A third aspect of risk related to rail operations lies in rail to rail collisions and derailments, some of which may be catastrophic. A discussion of safety programs implemented by the rail industry at national and international levels concludes the chapter.

INTRODUCTION

All transportation operations pose safety risks to the users of the transportation system, as well as to the general public; railway operations are no exception. The introduction provides a general overview of transportation safety in the United States in order to set the stage for a more detailed examination of America's only inter-city rail passenger operation – the National Railroad Passenger Corporation

DOI: 10.4018/978-1-5225-0102-2.ch006

(Amtrak). For the Introduction, data was obtained from the United States Department of Transportation's Bureau of Transportation Statistics (BTS) for three ten-year intervals of 1990, 2000 and 2010. The BTS provides a wide variety of data covering all modes of transport in the U.S.

In the United States in 2010, 32,999 persons were fatally injured on the highway system; of these, Figure 1 indicates 12,491 were passenger car occupants (BTS, 2014).

On the general railway network of the United States, 600 people were killed in 2010; of these, 134 were involved in collisions with Amtrak operated passenger trains or were casualties related to Amtrak operations (FRA, 2014). The general railway network of the United States, as defined by the Federal Railroad Administration (FRA), "refers to the network of standard gage track over which goods may be transported throughout the nation and passengers may travel between cities and within metropolitan and suburban areas" (CFR, 2014). The general railway network comprises about 140,000 route miles and is generally privately owned by seven major railroads, referred to as Class 1 railroads due their having annual revenue greater than $433.2 million, along with 21 regional and 510 local railroads (STB, 2014).

To round out the introductory summary, in 2010, 473 people were killed in general aviation incidents and 232 were killed in transit operations (bus, light rail, trolley) as reported to the Federal Transit Administration (FTA) as illustrated in Table 1 (BTS, 2014).

Table 1 indicates that since 1990, fatalities in the highway, transit and air modes have decreased significantly, yet fatalities due to Amtrak operations have remained

Figure 1. Highway Fatalities in the United States in 2010

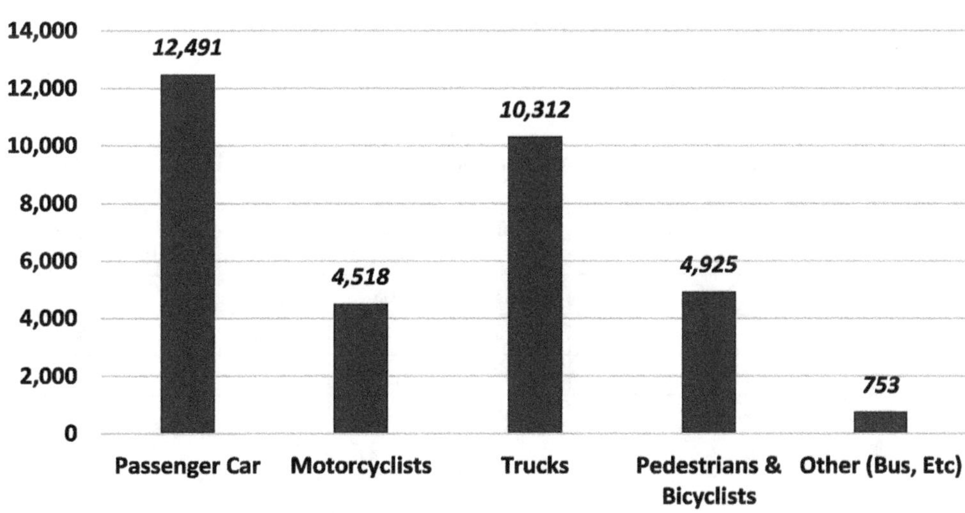

Table 1. Transportation Fatalities by Mode: 1990 to 2010

Transportation Fatalities	1990	2000	2010	Percent Change
Amtrak	127	135	134	5.5%
Transit	339	295	232	-31.6%
Air	866	764	473	-45.4%
Highway (Auto Passenger)	24,092	20,669	12,491	-48.2%

steady. Examination of the raw numbers of incidents is not generally considered appropriate when making comparisons between transportation modes and should be converted to rates. When the raw numbers of fatalities are converted into rates per 100,000 people in the United States, as illustrated in Table 2, passenger travel by Amtrak, is approximately 101 times safer than traveling on the highway system in an auto or other light duty passenger vehicle (4.04 /.04 = 101).

However, when safety performance is examined with a different metric, that being the number of fatalities per 100 million passenger miles of travel, Amtrak, does not perform as well other modes of transportation as shown in Table 3.

The remainder of the Chapter discusses safety aspects of passenger rail operations; including:

1. A review of rail safety responsibilities in the United States.
2. A discussion of rail safety performance between 2004 and 2013 comparing rail operations in general with those of Amtrak encompassing:
    a. Incidents involving passengers, employees, and trespassers not at highway-rail crossings
    b. Highway-rail crossing incidents
    c. Train to train related incidents

Table 2. Transportation Fatality Rates Per 100,000 Population by Mode: 1990 to 2010

Fatalities Per 100,000 Population	1990	2000	2010	Percent Change 1990 to 2010
Amtrak	0.05	0.05	0.04	-14.9%
Transit	0.14	0.10	0.08	-44.8%
Air	0.35	0.27	0.15	-56.0%
Highway (Auto Passenger)	9.66	7.34	4.04	-58.2%

Table 3. Transportation Fatality Rates Per 100 Million Passenger Miles of Travel by Mode: 1990 to 2010

Fatalities Per 100 Million Passenger Miles	1990	2000	2010	Percent Change 1990 to 2010
Amtrak	2.10	2.46	2.09	-0.5%
Transit	0.82	0.65	0.44	-46.5%
Air	0.25	0.15	0.08	-66.5%
Highway (Auto Passenger)	1.06	0.67	0.44	-58.5%

3. The Federal Railroad Administration's approach to ensuring safe high speed passenger rail transportation.
4. Discussion of national and international safety programs.

Railroad Safety Responsibility

Railroad safety in the United States is under the jurisdiction of the Federal Railroad Administration (FRA). FRA's mission, "*is to enable the safe, reliable, and efficient movement of people and goods for a strong America, now and in the future*" (FRA, 2014).

"FRA's Office of Railroad Safety promotes and regulates safety throughout the railroad industry. The office executes its regulatory and inspection responsibilities through a diverse staff of railroad safety experts. The staff includes 400 Federal safety inspectors who operate out of eight regional offices. Each regional administrator is supported by two deputy regional administrators, chief inspectors, supervisory specialists, grade crossing safety managers and safety inspectors for five of the safety disciplines focusing on compliance and enforcement in:

- Hazardous Materials
- Motive Power and Equipment
- Operating Practices
- Signal and Train Control
- Track" [(8)]

FRA's Office of Railroad Policy and Development is the program manager when it comes to developing and implementing high speed rail programs in the United States, primarily through its Office of Passenger and Freight Programs that administers grant funds provided to states to implement high speed rail programs.

"FRA's Office of Safety and Office of Railroad Policy and Development work together to implement a comprehensive strategy to eliminate risk on railroads. FRA accomplishes this by merging proven safety approaches - like our rail safety inspections and oversight programs - with performance-based safety measures that improve safety culture, harness technology and research, then apply the best of what we've learned to establish benchmarks, develop and ensure adherence to best practices, and implement new safety management programs" (FRA, 2014).

Railroad Safety Data

FRA maintains a wealth of railroad safety and system performance data reaching as far back as 1975. There are five principal FRA databases accessible by the public:

- 6180.54: Rail Equipment Accidents – includes train to train collisions, derailments, fires.
- 6180.55: Operational Data – includes train miles of operation, number of passengers carried and employee hours.
- 6180.55a: Railroad Illness and Injury – includes data covering injuries to users, employees and trespassers NOT at a highway-rail crossing.
- 6180.57: Highway-Rail Accidents – includes data for any collision between on-track rail equipment and a highway user that occurs within the confines of a roadway-railway intersection.
- 6180.71: National Inventory of Grade Crossings – complete inventory of all at-grade and grade separated crossings ever established.

The databases are easy to download and use and are available at the link below: http://safetydata.fra.dot.gov/OfficeofSafety/publicsite/downloads/downloads.aspx

A note concerning data processing: There is duplicative reporting between the Railroad Illness and Injury (6180.55a) database and the two databases covering Highway-Rail Accidents (6180.57) and Rail Equipment Accidents (6180.54) in that injuries reported on a 6180.57 or 6180.54 report must also be reported on the 6180.55a form.

Duplicate incidents in the Railroad Injury and Illness database are removed by using two fields; CAS57 and CAS54. If these fields are checked YES, then those records must be removed from the Injury and Illness database to avoid double counting of Highway-Rail Accidents and Rail Equipment Accidents. Likewise, the Rail Equipment Accident database also needs editing to remove records with a JOINTCD of 2 or 3 and TYPE of 07-Hwy-Rail Crossing to avoid double counting of Rail Equipment incidents.

Railroad Safety Performance: 2004 - 2013

This section provides an overall summary of railroad safety performance in the United States between 2004 and 2013 comparing the safety performance of Amtrak, as a specific subset of railroad carrier to all other rail carriers in the United States. The ten-year period was selected as it is the most recent ten-year period for which complete data is available when this Chapter was prepared.

The analysis focuses on Amtrak as Amtrak is the only provider of high speed passenger rail service in the United States and provides the only domestic benchmark of potential high speed passenger rail safety statistics. The summary addresses the users of the rail system and what happens to them (Injury and Illness), the interaction between trains and highway users at grade crossings (Highway-Rail Accidents), and on train-to-train incidents including collisions and derailments (Rail Equipment Accidents).

Railroad Injuries and Illnesses [6180.55a]: 2004 - 2013

Injuries and illnesses are reported by railroads on FRA Form 6180.55a. Included are ten sub-categories covering six types of employee or contractor, two types of authorized non-trespasser (e.g. patron at a station or depot); passengers on trains, and trespassers. Table 4 shows that there were a total of 4,806 railroad fatalities reported between January 1, 2004 and December 31, 2013 (NOT including fatalities

Table 4. Railroad Injuries and Illnesses: 2004 – 2013

Type of Person	USA: 2004-2013 (Excluding Amtrak)	Amtrak: 2004-2013	USA Total: 2004-2013
Worker on Duty	135	3	138
Employee Not on Duty	1	0	1
Passenger on Train	12	16	28
Non-trespasser on Property	85	13	98
Trespasser	3,782	712	4,494
Worker on Duty-Contractor	10	2	12
Contractor-Other	26	1	27
Worker on Duty-Volunteer	0	0	0
Volunteer-Other	0	1	1
Non-trespasser off Property	6	1	7
Total	**4,057**	**749**	**4,806**

Source: Federal Railroad Administration Railroad Injury and Illness Database

due to Highway-Rail or Rail Equipment Accidents which are covered separately later in the Chapter). Trespassers are the leading cause of railroad related deaths.

Measuring and quantifying the relative safety of one rail carrier against another is difficult. It is relatively simple to tabulate the number of reported incidents and related casualties. Converting raw numbers into rates normalizing the exposure to risk so that one rail carrier can be fairly compared against other carriers is a challenge. There are six potential risk denominators available in the Operational Data database:

- Employee hours worked.
- Passenger miles of transportation – 1 passenger carried for 1 mile = 1 passenger mile.
- Revenue passengers carried – number of passengers transported.
- Freight train miles of operation – number of train miles in freight service.
- Passenger train miles of operation – number of train miles in passenger service.
- Total train miles of operation: sum of passenger, freight, yard switching and other miles of operation.

No single performance measure equitably compares all rail carriers regardless of size, geography and type of traffic handled. The purpose of providing these measures within this analysis is to provide an initial point of discussion regarding the need to develop consistent and equitable rail safety performance measures for the future. Table 5 illustrates how Amtrak as a rail carrier (including all operations - commuter, regional and long-distance), compares against the rail industry as a whole.

Two measures; the number of incidents per million total train miles and the number of incidents per million employee hours provide a general performance

Table 5. Performance Comparison: All Railroads (excluding Amtrak) Against Only Amtrak

Performance Metrics of Number of Incidents Per Million	All RRs in USA (Excluding Amtrak)	Amtrak	Amtrak Outcome
Employee Hours	15.40	39.43	Poor
Passenger Miles	0.52	0.24	Good
Revenue Passengers Carried	11.06	53.17	Poor
Freight Train Miles	11.86	N/A	N/A
Passenger Train Miles	103.64	39.47	Good
Total Train Miles	9.24	37.48	Poor

comparison against all rail carriers while the remaining metrics compare just passenger carriers. In number of incidents per million passenger miles and number of incidents per million passenger train miles, Amtrak compares favorably to the universe of other passenger carriers; in this case about 20 commuter railroads along with a variety of tourist and museum operations. However, in the larger picture of railroad operational safety, Amtrak compares very poorly against all other railroads in the number of incidents per million total train miles and number of incidents per million employee hours.

Railroad Injuries and Illnesses: Focus on Railroad Related Trespassing

Trespassing is the leading cause of rail related fatalities. Trespassing is a difficult challenge for the rail industry to address, largely due to the fact that trespassing is often seen as a "local," or community based issue that the industry is not directly responsible for. Compounding the issue is the choice by some to use railroads as a means to end their lives. Suicide by train accounts for as many as half of all reported trespassing incidents (Savage, 2014).

Trespassing is a universal problem affecting railroads worldwide. In 2013, the International Union of Railways indicated that trespassing accounted for approximately 60 percent of rail related deaths as shown in Table 6 (UIC, 2014).

The International Union of Railways (UIC) has taken the lead in addressing rail related suicides. The UIC, in association with the European Railway Agency (ERA) and a number of other European rail entities, established the Reduction of Suicides and Trespass on Railway Property (RestRail) project. "The aim of the RestRail project is to reduce the occurrence of suicides and trespass on railway property, and the service disruption and other consequences these events cause by providing the

Table 6. 2013 Fatalities Reported by 21 Member Nations of the International Union of Railways in Europe

Type of Person	Fatalities	Percent
Trespassers	643	59.9%
Level Crossing Users	279	26.0%
Others	29	2.7%
Staff	27	2.5%
Passengers	96	8.9%
Total	**1,074**	**100.0%**

rail industry with analysis and identification of cost effective prevention and mitigation measures" (UIC/RestRail, 2014).

"RestRail's objectives are:

- Analyze the causes of suicides and trespass on railway property based on existing data, studies and research results.
- Identify the various available prevention and mitigation measures and analyze their conditions for success in the rail environment.
- Field test selected measures and evaluate their effectiveness.
- Propose a toolkit of the most relevant and cost-effective measures and recommendations at European level both to reduce the occurrence of incidents as well as to manage incidents and reduce consequences."

RestRail recently wrapped up its effort with the release of the Tool Box which is available at: http://www.restrail.eu/toolbox/.

Highway-Rail Accidents [6180.57]: 2004 - 2013

Much attention is focused on the interface where highway meets rail at grade (level) crossings. FRA and the Federal Highway Administration provide over $200 million dollars annually to improve safety at grade crossings. In addition, many states, such as Illinois, provide additional funds to eliminate risks at level crossings. Grade separations are the ideal improvement, yet can easily cost $20 million for a relatively simple urban separation of a single track railroad with a two lane highway.

In 2004, there were a total of 3,085 collisions between trains and highway users at grade crossings resulting in 371 fatalities and 1,095 injuries. By 2013, those numbers had deceased by over 30 percent to 2,089 collisions with 289 fatalities and 957 injuries. The steady decrease over time yields one of transportation safety's greatest success stories as shown in Figure 2.

In comparing Amtrak against the rest of the operating railroads in the United States between 2004 and 2013, a total of 23,131 collisions occurred of which 1,276 involved Amtrak as the primary reporting railroad involved. It is important to note that when collisions at highway-rail grade crossings occur, the collision can be assigned to either the railroad that was operating the train involved in the collision, or the railroad that owned the crossing and responsible for maintenance of the crossing surface, warning devices, etc. Amtrak's highway-rail grade crossing collision rate per million train miles of operation compares very favorably against other railroads in the United States.

Figure 2. Highway-Rail Grade (Level) crossing Collisions: 2004 - 2013

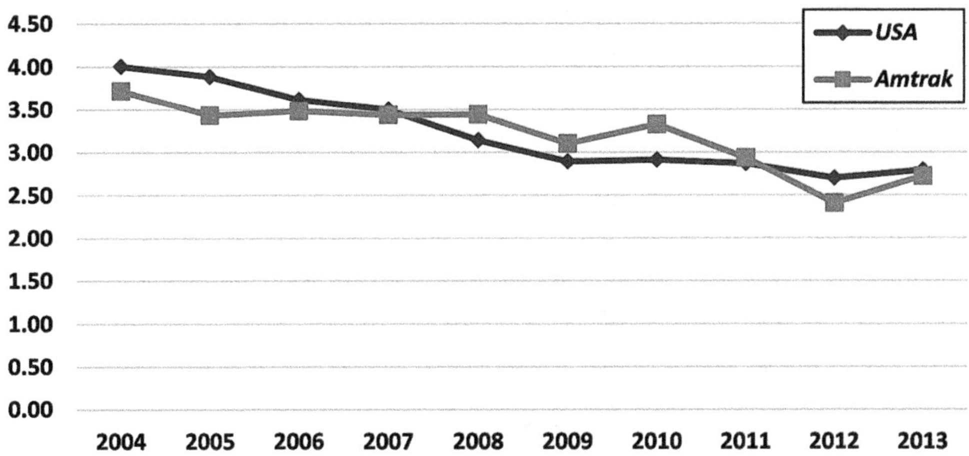

Rail Equipment Accidents [6180.54]: 2004 - 2013

Train accidents and incidents include all types of train to train collisions, such as head-on and side-swipe collisions, as well as derailments, fires and other types of train accidents, such as a train striking an obstruction on the tracks, but in this analysis, do NOT include highway-rail crossing incidents. Between 2004 and 2013, train accidents decreased by 45.7 percent of all railroads excluding Amtrak, and by 45.9 percent for Amtrak as shown in Figure 3. Over the ten-year period, Amtrak actually

Figure 3. Rail Equipment Accidents: 2004 - 2013

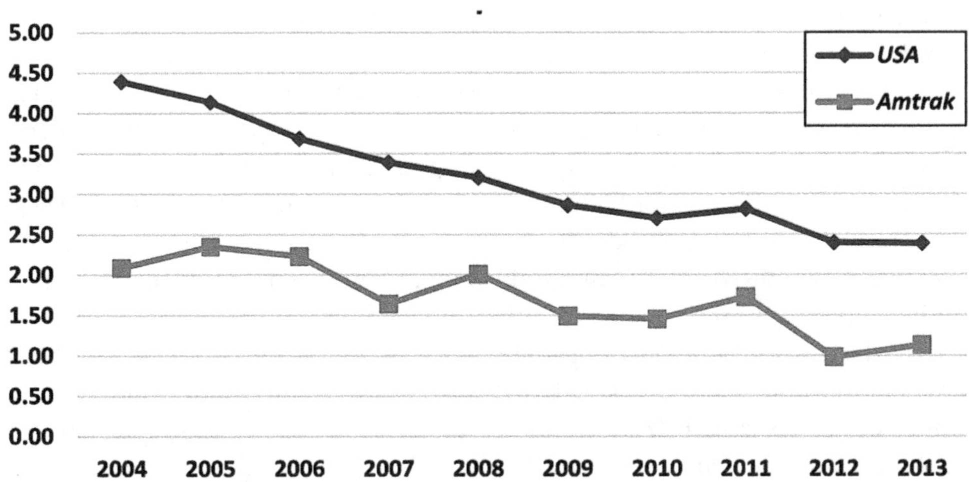

has a much better train accident rate per million train miles of 1.69 compared to the rate for all other railroads of 3.31 train accidents per million train miles.

Fortunately, train accidents where higher speed inter-city passenger trains have derailed or collided with other trains, are very rare. However when they do occur, as in Germany (1998), Spain (2013) and China (2011), they can lead to a catastrophic loss of life.

Table 4 and Table 5 illustrated that Amtrak has a higher rate of fatalities for system users, such as passengers on trains and at stations, as well as employees and trespassers, than the rail industry as a whole. However, Amtrak has a safer rate of incidents at highway-rail grade crossings and for train to train and other related train accidents as shown in Figures 2 and 3.

The Federal Railroad Administration Approach to High Speed Rail Passenger Safety

The FRA focuses on a systems based approach to safety based on risk analysis and risk mitigation. No transportation system is perfectly safe; FRA's safety objectives are to avoid risk when possible and to minimize the consequences when incidents do occur. FRA employs a three-step process to ensure safety of high speed passenger rail operation:

1. Establish safety standards and program guidance,
2. Apply a system safety approach to address safety concerns on specific rail lines, and
3. Ensure that railroads involved in passenger train operations can effectively and efficiently manage train emergencies.

FRA endeavors to achieve uniformly safe rail passenger service, regardless of speed. Since the severity of collisions and derailments increases with speed, safety performance targets for preventive measures are tiered to become more stringent as speed increases. Current FRA regulations include Tier I equipment safety standards for passenger trains operating at speeds of up to 125 mph. FRA established additional standards (Tier II) for high-speed trains that operate up to 150 mph. These Tier II requirements address crashworthiness, crash energy management, vehicle suspension systems, brake systems, train configurations and other elements critical for high-speed train sets.

The Tier II standards only impact the Amtrak Acela Service on the Northeast Corridor because Acela is currently the only Tier II application operating in the United States. Existing FRA track safety standards are comprehensive and specify track geometry and cant deficiency for FRA Classes 1 through 9 at speeds up to

220 mph. The track standards include requirements for Class 6 track (110 mph) and above for remote monitoring of car body acceleration and truck hunting, and equipment qualification on actual routes. FRA's proposed regulatory framework to guide the implementation of high speed rail service is shown in Table 7 (FRA, 2010).

Table 7. FRA Regulatory Framework for High Speed Rail Implementation as of January 2011

System Component	0 to 79 mph Conventional	80 to 110 mph Emerging HSR	111 to 125 mph HSR Regional	126 to 150/160 mph Mixed Ops	151 to 200+ mph HSR Dedicated
Track	Class 4	Class 5 & 6	Class 7	Class 8	Class 9
Rail Traffic Mix	Mix passenger & freight	Mix passenger & freight	Mix passenger & freight	Mix passenger & freight	Passenger Only
Passenger Equipment	Conventional	Tier 1	Tier 1	Tier 2	Guidance TBD
Passenger Occupied Vehicle in Lead	Permitted	Permitted	Permitted	Prohibited	Guidance TBD
Signal & Train Control	Traffic control or PTC	PTC vitality & perimeter control above 90 mph	PTC vitality & perimeter control above 90 mph	PTC incursion & R-O-W safety strategy integrated	PTC – system safety plan integrated
Private Hwy/Rail Grade Xings	Automated warning of locked gate	Automated warning of locked gate & dispatch control	Barriers above 110 mph; presence detection tied to PTC over 110 mph	Prohibited	Prohibited
Public Hwy/Rail Grade Xings	Automated warning; supplemental safety measures where warranted	Sealed corridor	Barriers above 110 mph; presence detection tied to PTC over 110 mph	Prohibited	Prohibited
Maint. Of Way Safety Mgmnt.	Address within SSP framework	Guidance TBD	Guidance TBD	Guidance TBD	Guidance TBD
Right of Way Incursion Prevention	System safety program / collision hazard analysis	Guidance TBD	Guidance TBD	Required CFR 236.362	Required CFR 236.362
System Safety Program	Guidance TBD	Guidance TBD	Guidance TBD	CFR Subpart G, Part 238	Guidance TBD

Source. Federal Railroad Administration

FRA's safety strategies are designed to prevent incidents from occurring in the first place and mitigating the consequences of an incident when one does occur through development and implementation of emergency response plans, as well as system safety plans. Outlined below are FRA's safety strategies:

1. Prevention Strategies.
    a. Vehicle-Track Interaction. Track Safety Standards (49 CFR Part 213) specify requirements for nine classes of track based on maximum allowable speed as shown below. The standards for higher speeds focus on vehicle-track interaction, as well as track geometry, rail integrity and appropriate inspection practices.
    b. Positive train control.
        i. Must be installed by December 31, 2015 on lines where passenger service is regularly operated.
        ii. Demonstrate that opportunities for incursion into the right-of-way (e.g., vandalism, motor vehicles deviating from adjacent roadways) have been addressed through prevention and/or detection and warning.
    c. Grade crossing safety.
        i. Promote "sealed corridors."
        ii. Eliminate all redundant or un-necessary crossings.
        iii. Install the most sophisticated traffic control/warning devices compatible with the location (e.g., median barriers, special signage (possible active advanced warning), four-quadrant gates) where train-operating speeds are between 80 and 110 mph.
        iv. Protect rail movement with full width barriers capable of absorbing the impact of highway vehicles where train-operating speeds are between 111 and 125 mph.
        v. Eliminate or grade-separate all crossings where trains travel at speeds above 125 mph.
    d. Maintenance-of-way safety management: Five principles of Roadway worker protection systems - 49 CFR Part 214.
        i. A person who is not fouling a track will not be struck by a train.
        ii. A person who is fouling a track upon which a train will not move will not be struck by a train.
        iii. No person should foul a track unless that person knows either that no train will arrive or the person on the track will be able to move to a place of safety before a train arrives.
        iv. Each roadway worker bears the ultimate responsibility for his own on-track safety.

    v. Each employer is responsible for providing the means for achieving on-track safety to each roadway worker employee.
  e. Right-of-way safety. 49 CFR Part 213.361 of the Track Safety Standards requires for Class 8 and 9 track, the track owner must submit for approval a "right-of-way plan" for the prevention of:
    i. Vandalism.
    ii. Launching of objects from overhead bridges or structures into the path of trains.
    iii. Intrusion of vehicles from adjacent rights of way.
2. Mitigation.
    a. Structural standards for infrastructure components to minimize consequences.
    b. Passenger equipment safety standards for equipment to minimize injuries.
3. Emergency Management.
    a. Standards for the preparation, adoption and implementation of emergency preparedness plans under 49 CFR Part 239, Passenger Train Emergency Preparedness.
4. System Safety Programs that identify safety needs and manage them over time, including hazard identification, which focuses on opportunities for risk reduction in the particular circumstances of the specific passenger railroad. Programs must;
    a. Be defined and documented.
    b. Include hazard management processes designed to proactively identify, assess and mitigate hazards before accidents occur.
    c. Be fully implemented by the passenger railroad.
    d. Be audited for compliance.
5. Real-Time monitoring. A variety of technologies are now available to monitor the health and performance of the railroad operating system in real time. Examples include onboard sensors, wayside detection devices and autonomous track geometry systems.
    a. Motive power and equipment monitoring – accelerometer applications.
    b. Right-of-way systems – rail integrity, obstruction detection systems.
    c. Level crossing health and remote monitor systems – four-quadrant gate vehicle detection systems.

FRA's objective is to minimize risk by employing a systems based approach identifying potential sources of risk and then mitigating the risk to the greatest extent possible. This approach has worked very well in reducing the occurrence of all type of rail incidents and minimizing the consequences and impacts of the incidents that do occur. FRA has more rule makings in the works to further reduce both the likelihood of an incident as well as the severity of any future incident.

Amtrak Higher Speed Performance

As demonstrated earlier, Amtrak's safety performance in regard to casualties incurred is less satisfactory than that of the industry as a whole. Amtrak's casualties were primarily due to trespassers (712 of 749 total deaths or 95%). In contrast, Amtrak's safety performance in regard to highway-rail crossing safety and train accident / incident safety is as good as if not better than the industry as a whole.

Ideally, this chapter would be able to shed light on Amtrak's higher speed operations. Unfortunately, the FRA data available do not provide a means to distinguish Amtrak commuter operations from long distance train operation from Acela operation from non-Acela high speed operation. What the data can provide is a simple analysis of reported incidents where train speeds were equal to or exceeded 79 mph.

Between 2004 and 2013, Amtrak trains were involved in 884 rail equipment accidents that did not involve highway-rail crossings. Of the 884 incidents, 141 occurred at a train speed of 79 mph or greater (15.9%). Figure 4 provides a breakdown of the type of rail equipment accident occurring at train speeds of 79 mph or greater. Pantograph damage was the leading cause of train incidents accounting for 70 percent of high speed train accidents.

Figure 4. 141 Amtrak Rail Equipment Accidents at 79 mph or Greater: 2004 - 2013

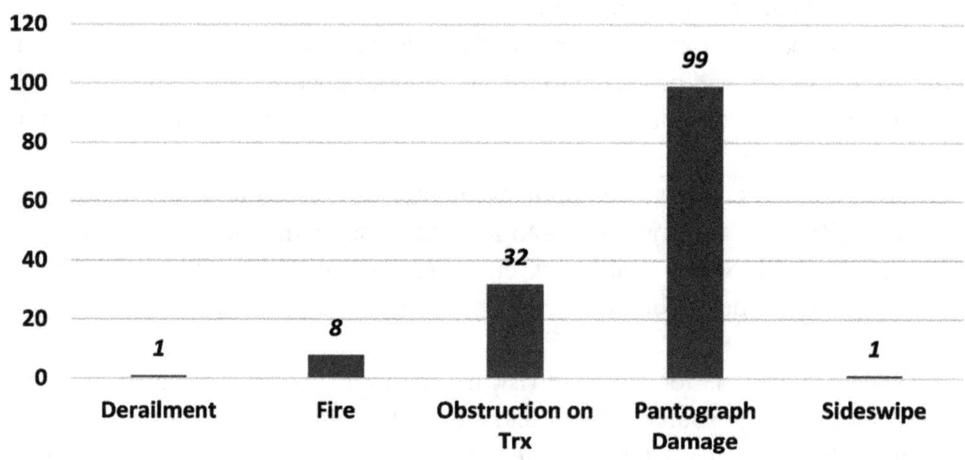

Safety Is Good Business

Finally, this Chapter would not be complete without a discussion regarding the rail industry's programs to address safety issues. The rail industry has long embraced the traditional 3E program of Education, Engineering and Enforcement of Traffic Laws to optimize its safety initiatives. The 3E program is best personified by Operation Lifesaver (OL). OL was founded in 1972 by the Union Pacific Railroad in Idaho and has since spread to all 50 states, Canada and several countries overseas, including Estonia and Argentina.

Education is fundamental to everything we do at Operation Lifesaver. Our rail safety programs offered by dedicated volunteer Presenters in schools, civic and community organizations along with specialized training for law enforcement, professional drivers and emergency first responders give people information they need to stay safe around railroad property and rights-of-way. Our programs provide valuable facts about trains, illustrate how drivers can safely navigate highway-rail grade crossings, and reinforce that it's illegal and unsafe to ever walk on, or use railroad tracks for, recreation.

The challenge for engineering professionals is to engineer out as many decisions for the motorist as possible," says one railroad safety expert. While Operation Lifesaver's role is primarily to promote education, engineering is one of the important 3E's of rail traffic safety. We support and encourage the kinds of continued engineering research and innovation responsible for having greatly improved the safety of America's railroad crossings because whenever two modes of transportation share the same section of roadway, as they do at highway-rail grade crossings, the possibility for a collision exists. Engineering technologies play a critical role in helping minimize and reduce those collision risks.

Operation Lifesaver's close partnership with the law enforcement community throughout America has strengthened and improved public safety. It is critical that when highway-rail intersection collisions do occur, law enforcement officers--often the first responders--are familiar with both railroad operations and highway-rail grade crossing conditions. Operation Lifesaver offers a special course teaching officers how to ensure their personal safety, both while responding to rail collision incidents and throughout the investigation. (Operation Lifesaver, 2014)

Analysis conducted by Northwestern University in 2001 "concluded that about two-fifths of the decrease (in collisions at Level Crossings) is due to factors such as reduced drunk driving and improved emergency medical response that have improved safety on all parts of the highway network. The installation of gates and/ or flashing lights accounts for about a fifth of the reduction. The development in the 1970s and early 1980s of the Operation Lifesaver public education campaign, and the installation of additional lights on locomotives in the mid 1990s, each led to about a seventh of the reduction. Finally, about a tenth is due to closure of crossings resulting from line abandonments or consolidation of little-used crossings" (Savage & Mok, 2005).

Individual railroads maintain extensive public safety information on their corporate websites. BNSF Railway, for example has several pages dealing with grade crossing safety, establishment of quiet zones, maintenance of level crossings and the BNSF crossing closure program. BNSF's crossing closure program has led to the closure of several thousand public and private highway and pedestrian pathway level crossings. BNSF's information can be located at: http://www.bnsf.com/communities/pdf/4_GradeCrossingSafety.pdf (BNSF Railway, 2014).

A third example of the rail industry's commitment to safety is best exemplified by the initiation of International Level Crossing Awareness Day (ILCAD). ILCAD "is a joint commitment continuing from the success of the first European Level Crossing Awareness Day held on 25th June 2009 in 28 countries raising public awareness on the dangers of misbehavior at level crossings. Level crossing accidents account for only 1.2% of road deaths but 29% of all rail fatalities based on European statistics. Road and rail organisations from participating countries acknowledged their shared responsibility to deal with this issue by organizing safety events to develop public awareness and safe behaviour at and around level crossings. Additionally, many of them have signed the European Road Safety Charter which is an initiative of the European Union.

The Charter has the objective of developing concrete actions, assessing results and further heightening awareness about the need to reduce traffic accidents and fatalities through the exchange of best practices in traffic safety in Europe, while adding a European dimension to individual or collective initiatives. Due to the success of the first European Level Crossing Awareness Campaign (ELCAD) at national, European and international levels in 2009 it has been agreed that the fourth week of June in the years to come The members particularly welcomed the active participation of Argentina and Israel amongst non-European partners in ELCAD 2009 and the support of Canada, South Africa and the United States and look forward further develop joint actions for the next event giving it a truly international dimension" (UIC/ILCAD, 2014). ILCAD for 2015 will be held on June 3.

SUMMARY

How safety is measured influences how safety programs are designed and implemented. The introduction presented serval ways to measure safety among the four principal modes of transport: highway, air, transit, and Amtrak. Depending upon the metric used, Amtrak is either 101 times safer than highway travel (incidents per 100,000 of population), or 4.75 times more lethal than highway travel (per 100 million miles of passenger travel). Regardless of how safety is measured, the discussion of rail related casualties demonstrated that the vast majority of fatalities related to inter-city rail transport arise from trespassing.

Trespassing is a difficult challenge for the rail industry to address, especially since as much as half of all trespassing incidents are due to intentional efforts by trespassers to end their lives by stepping in front of a train. Higher speed, scheduled trains are particularly affected by rail related suicide.

The second leading cause of rail related fatalities are level crossing incidents. Contrary to how high-speed rail has been developed internationally, higher speed operations implemented in the United States include the presence of level crossings. Fortunately, traditional education, engineering and traffic law enforcement programs have been very successful at reducing the number of collisions between trains and highway users at level crossings.

The third, but perhaps most consequential area of risk arises from rail equipment accidents. Rail equipment accidents where higher speed inter-city passenger trains have derailed or collided with other trains, are very rare. However when they do occur, as in Germany, Spain and China, they can lead to a catastrophic loss of life.

High speed passenger rail safety in the United States is the responsibility of the Federal Railroad Administration and has been very thoughtfully addressed. Amtrak's role as the current provider of high speed passenger rail service appears to be working fine with minimal fatalities and no major incidents to date. The implementation of new high (and higher) speed passenger rail services will proceed hand-in-hand with the implementation of new rules and standards for infrastructure, rolling stock and motive power, thus keeping the traveling public safe.

Opportunities for future research lie primarily with being able to further analyze Amtrak's current experience with higher speed operations. A very fertile field of future research also lies in documenting the best safety practices of high speed rail operators throughout the world. At present, there is very little literature available to identify best safety practices internationally and transfer those practices to domestic high speed rail operations.

REFERENCES

International Union of Railways (UIC). (2014). RESTRAIL. Seventh Framework Programme for Research and Technological Development. Retrieved from http://www.restrail.eu/

International Union of Railways (UIC). (2014). Safety Database Report Significant Accidents occurred in Europe during the year 2013 Public Report [Data file]. Retrieved from http://safetydb.uic.org/IMG/pdf/Public_Report_2014.pdf

International Union of Railways (UIC). (2014). International level Crossing Awareness Day (ILCAD). Retrieved from http://www.ilcad.org/History.html

Operation Lifesaver. (2014). Mission Statement. Retrieved from http://oli.org/about-us

BNSF Railway. (2014). Safety and Security. Retrieved from http://www.bnsf.com/communities/safety-and-security/railroad-grade-crossings/

Savage, I. (2014) Analysis of Fatal Train-Pedestrian Collisions in Metropolitan Chicago 2004-2012. *Proceedings of the 2014 Global Level Crossing and Trespass Prevention Symposium*. University of Illinois at Urbana-Champaign.

Savage, I., & Mok, S. (2005, August). Why has Safety Improved at Rail-Highway Grade Crossings? *Risk Analysis*, *25*(4), 867–881. doi:10.1111/j.1539-6924.2005.00642.x PMID:16268935

U.S. Department of Transportation. Federal Railroad Administration – Office of Safety. (2009) FRA High-Speed Passenger Rail Safety Strategy; Retrieved from https://www.fra.dot.gov/eLib/Details/L03624

U.S. Department of Transportation. Bureau of Transportation Statistics. (2014). National Transportation Statistics, Chapter 2 – General Aviation Fatalities in 2010 [Data file]. Retrieved from http://www.rita.dot.gov/bts/sites/rita.dot.gov.bts/files/publications/national_transportation_statistics/index.html#chapter_2

U.S. Department of Transportation. Bureau of Transportation Statistics. (2014). National Transportation Statistics, Chapter 2 – Transit Fatalities in 2010 [Data file]. Retrieved from http://www.rita.dot.gov/bts/sites/rita.dot.gov.bts/files/publications/national_transportation_statistics/index.html#chapter_2

U.S. Department of Transportation. Bureau of Transportation Statistics. (2014). National Transportation Statistics, Chapter 2 – Auto Fatalities in 2010 [Data File]. Retrieved from http://www.rita.dot.gov/bts/sites/rita.dot.gov.bts/files/publications/national_transportation_statistics/index.html#chapter_2

U.S. Department of Transportation. Federal Railroad Administration. (2014). Mission statement. Retrieved from http://www.fra.dot.gov/Page/P0002

U.S. Department of Transportation. Federal Railroad Administration. (2014). Rail passenger / Amtrak Fatalities in 2010 [Data file]. Retrieved from http://safetydata.fra.dot.gov/OfficeofSafety/publicsite/on_the_fly_download.aspx

U.S. Department of Transportation. Federal Railroad Administration. (2014) 49 CFR Ch. II Part 209 Appendix A. Statement of Agency Policy Concerning Enforcement of the Federal Railroad Safety Laws. Washington, DC: U.S. Government Printing Office.

U.S. Department of Transportation. Federal Railroad Administration – Office of Safety. (2014). Retrieved from http://www.fra.dot.gov/Page/P0010

U.S. Department of Transportation. Surface Transportation Board. (2014). Definition of Class 1, 2 and 3 railroads. Retrieved from http://www.stb.dot.gov/stb/faqs.html

Chapter 7
Exploring the Airline–High Speed Rail Collaboration Model:
Efficient Services and Mutual Benefits

Peggy Daniels Lee
Kelley School of Business, Indiana University, USA

George VandeWerken
Providence Bank, USA

Raj Selladurai
Indiana University Northwest, USA

ABSTRACT

Studies have shown that short-haul airline passenger traffic (less than 500 miles) is decreasing nationwide. This decline may be attributed to legacy airlines' rising costs (especially fuel), increased airport congestion, and increased travel time due to post-September 2001 TSA security screening. Previous studies tended to look at the substitution of high-speed rail and other transportation modes for air travel, especially as travel times shorten. Substitution usually takes the form of collaboration or competition between competing modes and competing carriers. This chapter presents an alternative view – with a discussion of the proposition that air carriers may benefit more from collaborative arrangements that allow them to "own" at least a portion of the intermodal passenger experience rather than shifting or

DOI: 10.4018/978-1-5225-0102-2.ch007

transferring passengers to competing non-air modes. As we believe that this proposal has merit and is worth consideration, we conclude the chapter with a research agenda designed to empirically test the proposition.

INTRODUCTION

Domestic airline travel distance is defined by flight time, which can be translated into distance. Domestic air carriers commonly define a short haul flight as one that takes less than one and a half hours to complete. This roughly translates into 500 miles or less. However, some authors use the 400-mile-or-less criterion. There has been significant anecdotal evidence that short haul air travel is declining nationwide. Murphy and Meilus (2012) empirically supported these assertions. They found that passengers traveling less than 400 miles decreased by 26 percent since the mid-1990's. The number of air travel passengers flying distances between 1,000 and 2,500 miles increased by 50 percent during the same time period. Elking and Windle (2014) showed that the number of domestic US air passengers traveling less than 500 miles declined by 17 percent over the 1995-2010 time period.

At the same time, high-speed rail has increased as the travel time has decreased. For example, the two-hour rail trip between Paris and Lyon held a 60% share of the rail/air market when the trip took three hours. It is currently 90%. High-speed rail between Madrid and Seville captured 80% market share. Amtrak's Acela increased its market share between New York and Washington, D.C. from 37% to 75% between 2000 and 2011, and between New York and Boston from 20% to 54% during the same time period (Nixon, 2012). Further, air quality measured in carbon dioxide (CO_2) emissions rates improves with high-speed rail. Hu, et al. (2014) showed that high-speed rail emissions rates in the Northeast Corridor of the US (the Boston to New York route) are 23% of emission rates from regional jets and 41% of emission rates from automobiles. These few examples suggest that high-speed rail is a viable substitute for short haul air travelers.

Airlines can approach this trend by shifting their focus and investment to longer hauls, by trying to improve the short haul travel experience to avoid passenger attrition, by working to reduce costs, or by some combination of these and other factors. Clearly, as can be seen in the Amtrak Acela example, air-carrier passenger traffic and revenue have declined as passengers have shifted to rail. Therefore, the question deserves to be asked whether there may be another financially attractive alternative for the airlines to pursue that allows them to maintain short haul passengers and increase their revenues. Could airlines alter their business model to integrate the use of other modes to move passengers and baggage in ways that would provide benefits for both the airline industry and the passengers? In particular, could some form of passenger rail service operated in part by the airlines be used as a tool to integrate the movement of passengers over some short-haul airport routes such as Chicago and Indianapolis for example? And could this integration model generate significant additional profits for the airlines if it were done properly?

AIRLINE BUSINESS MODEL

The importance of the airline industry to the global economy cannot be overstated, particularly the US airline industry. In 2006, over 100 passenger airlines operated over 11 million flight departures per year, and carried over one-third of the world's air traffic (MIT, 2015). US airlines reported over $160 billion in revenues, with approximately 454,000 employees and over 8,000 aircraft operating 31,000 flights per day (MIT, 2015). Since 2006, fuel costs have been the largest airline expense, contributing to a general depression and reorganization of the industry in favor of low cost carriers, especially on short haul flights. Airport capacity has also emerged as a major challenge to the success of the legacy airlines. The only way to increase airport capacity is to build more runways and add more gates. However, the approval process for new airports/runways can be cumbersome and long because the airlines do not own the airports. In North America, Asia, and Europe, airports are owned and operated by airport authorities, which are quasi-government agencies. In the face of these capacity constraints, airport authorities have resorted to demand management policies such as slot restrictions, congestion pricing, and even auctioning access to major airports. Airlines pay the airport authorities for access and use of the airport space, which contributes to their cost issues. Therefore, the airlines have resorted to taxing airline passengers directly for everything from food on board to baggage handling. According to a report of the MIT Global Airline Industry Program (2015), "various taxes and fees for infrastructure support and security currently increase the cost of the average domestic airline ticket in the United States by about 16%" (page 7). These challenges in the current business model may continue to restrict airline revenue growth, the profitability of the airlines, the safety and security of passengers, and sustaining a state-of-the-art air transport infrastructure. Airlines may likely continue to see passenger erosion unless they find alternative ways to move large numbers of passengers from planes to some other mode, which is fully integrated into the rest of the air passenger traffic movement. A business model that actively embraces a multi-mode passenger movement strategy would look for ways to move passengers from planes to some fully integrated mode of air passenger traffic movement; for example, particularly where travel times may be comparable, and congestion relief significant. Doing so could be one means of reducing the pressure on the runway and airport gate infrastructure and improving airline profitability at the same time.

We use the corridor between Indianapolis, Indiana and Chicago, Illinois as a potential subject of a feasibility study. A recent examination of airline flight schedules between Indianapolis and Chicago indicates 17 flights per day going each way between Indianapolis and Chicago's O'Hare airport (Indianapolis International Airport, 2015). If we assume that each flight carries 100 passengers, then all the carriers combined move about 1,700 passengers per *day* each way or about 3,400 in total between Indianapolis and Chicago, which is about 51,000 passengers per month each way. By comparison, current Amtrak rail passenger traffic between these two cities is estimated to be considerably less. Since data that indicates rail passenger trips between these two destinations is not available, this estimate is based on a review of the annual station passenger data for Indianapolis (33,033 in 2014) and Lafayette (23,609 in 2014) combined. Based on this limited data it is clear that Amtrak may be carrying at most 4,000 passengers per month between Chicago and Indianapolis – and likely about 2,000 per month in each direction. Greyhound and MegaBus provide bus service between Chicago and Indianapolis and together average 12 to 13 trips per day each way. If we assume that 40 passengers travel on each departure, total passenger movement may be estimated at about 15,000 per month each way (40 x 13 x 30 = 15,600) (Greyhound, 2016) and (MegaBus, 2016). All modes of public transportation thus are carrying about 67,000 passengers each way each month.

By contrast, the most significant people movement competitor appears to be the interstate highway system, primarily Interstate 65, which runs from Indianapolis to Northwest Indiana, where traffic is routed onto Interstate 80/94 and then on to destinations throughout the Chicago metropolitan area. The location of the lowest average traffic count on Interstate 65 is just under 30,000 vehicles per day (29,898 just south of Rensselaer) (Indiana Department of Transportation, 2015).

If we assume that 1/3 of those vehicles are trucks, and of the remaining 20,000 vehicles approximately 1/3 of those 20,000 vehicles represent trips between the Indianapolis and Chicago areas, then auto passenger travel between these two metropolitan areas is likely to carry 10,000 people per day, if we further assume an average of 1.5 passengers per vehicle. This equates to about 300,000 per month and is at least five to six times more people than all other modes combined. (Note: these are rough estimates developed from limited data and from what seem to be reasonable assumptions, but they could be strengthened with better data, when available, showing actual passenger counts using each mode between these two destinations).

The most significant airline business opportunity here lies in developing ways to attract travelers from the interstate highways. Apart from auto travel, the airlines collectively move more passengers than any other mode of mass transit, and the prospect of moving more passengers between these two cities could provide significant additional revenue opportunities. For example, if the number of departures each way between Indianapolis and Chicago could be increased from 17 to 34, and the convenience and comfort of going in either direction improved, the potential revenue growth and profit from this one short haul route example could double. The airlines may need to consider whether their business model should be amended or expanded to include better integration of multi-modal movement of their passengers.

Admittedly, few models for this multi-modal airline business model currently exist. The most commonly found model presenting itself when conducting a cursory examination of the literature is collaboration between competing mode operators who agree to code sharing arrangements. One European example between Lufthansa and Deutsche Bahn allows passengers to include in their travel plans a rail segment between the Frankfort airport and the Stuttgart or Cologne train stations served by Deutsche Bahn (Global Air Rail Awards, 2012).

A Chinese example appeared in 2013 when several carriers announced plans to develop or expand partnerships with Chinese rail service (Hong Kong Economic Journal, 2016). Of course, the non-collaborative history of passenger movement on the Northeast corridor indicates a significant shift of passengers from air carriers to Amtrak, the sole provider of rail service offering competitive schedules and fares. However, none of these examples involves the airlines actually providing their service over rail but rather cooperating or collaborating with rail competitors – or simply turning over passengers to a competitor

THE CHALLENGES

Therefore, if the airlines where to adopt such a multi-modal business model, what might it look like and what obstacles would they have to overcome? If it were easy to do, why have not good examples emerged?

The answer is that becoming a multi-modal carrier involves major challenges, which is probably why US based carriers have not tried it. The lack of experience in operating in a rail environment, the difficulty of securing arrangements to operate on freight carrier rights-of-way, and the significant capital costs to develop and maintain a rail operation – these are all factors which doubtless discourage entry. However, US based airlines may need to look more closely at one of their counterparts in Europe, Virgin Airlines, which also operates Virgin Trains and manages both successfully (Milmo, 2015).

Closer to home, however, there is a recent innovative example of a rail service delivery alternative that might provide a different conceptual model for airline entry into the rail mode. Iowa Pacific Holdings, through its various subsidiaries, is a rail carrier that acquires and operates railroads (Boomberg Business, 2015). During 2014 – 2015, Iowa Pacific has been in negotiations with the Indiana Department of

Transportation (INDOT) and Amtrak to explore ways to expand and improve service on the dying Hoosier State route, currently an Amtrak-operated train providing four round trips per week between Indianapolis and Chicago. The State of Indiana together with several municipalities along the Hoosier State route subsidized this mediocre level of service with an annual operating appropriation of $2.7 million last year.

Both the supporters and detractors of this service's continuation have brought considerable pressure on INDOT. The interesting aspect of this story for our purposes, however, is that INDOT has found in Iowa Pacific a potential rail service partner to provide improved service – a partner that sees potential to transform service between Indianapolis and Chicago from a money-losing, inadequate operation to a modern-day, attractive travel alternative. The objective is to improve the quality of the travel experience and to move that service level to 12 round trips per day. The negotiations indicate the potential when new parties are introduced. "Under the proposed service, Amtrak would have served as the primary operator, working with host railroads, providing train and engine crews, and managing reservation and ticketing. Iowa Pacific would have provided the train equipment, train maintenance, on-board services and marketing" (AASHTO Journal, March 13, 2015). Iowa Pacific sees the potential to attract significant ridership if the frequency and quality of service are improved.

Although Iowa Pacific is not an airline, the arrangements it has negotiated in cooperation with Amtrak and INDOT could serve as an example for airline involvement. Rail transit involves a number of elements, and the proposed arrangement between Iowa Pacific, Amtrak, and INDOT offers one possible way to configure the elements between collaborating partners. What Iowa Pacific brings to this equation is its experience and expertise in attracting passengers, and increased passenger count would increase revenues and leverage high infrastructure costs much more effectively. For purposes of our discussion here, the question to be asked is what added potential exists to drive even greater numbers of passengers to this same rail corridor if airlines could provide some of the same elements proposed by Iowa Pacific? What kind of passenger traffic could airlines develop if they provided equipment, maintenance, on-board services, marketing, and, in addition, their own integrated ticketing and reservations services as well as baggage handling for rail service between the Indianapolis and Chicago airports? Additionally, what would happen to airline passenger traffic counts if this service could be incorporated into airline terminals at both ends?

The distance between O'Hare and the Indianapolis airport is just over 200 miles, and (for purposes of stimulating our thinking) between Indianapolis and Cincinnati is another 120 miles. If rail speeds could be improved to average 100 MPH, for example, the travel times would be just over an hour between Cincinnati and Indianapolis and just over 2 hours between Indianapolis and Chicago. Improved speed coupled with greater departure frequency and reliability of service would make rail travel between these cities very attractive. The travel times for these relatively short distances would compare very favorably with current air or auto travel.

Of course, this proposition needs to be tested. The revenue potential analysis for this type of enhanced service would require estimating how many additional travelers the airlines could attract if they offered quality rail service between these major airports – and included three additional stops. The Indianapolis to Chicago segment, for example, could attract additional riders if stops were to be made in Lafayette, Gary, and Chicago. These three additional stops would likely add 15 or 20 minutes travel time for the entire trip, but the airlines would be attracting passengers who might want to travel between Lafayette and Chicago, for example, or passengers who wanted to go from Indianapolis to downtown Chicago. These added riders would be attracted by the frequency and convenience of service, and they represent significant revenue opportunities that the airlines currently may be losing.

Airline entry into rail passenger movement would represent the introduction of a new element of discipline into the passenger movement business. As some commentators have pointed out, the amazing success of the US freight railroads has helped contribute to the demise of good passenger service in the US (Yglesias, 2015). For the most part passenger trains operate on the same infrastructure that supports the freight railroads, and passenger service is not the priority of the freight railroads. Currently, public policy has placed the entire burden of developing and expanding passenger rail service in this country on Amtrak. Anyone familiar with Amtrak's transportation mandate is aware of the political challenges Amtrak faces in obtaining funding and fighting to expand service. In fact, as one writer has suggested, a significant segment of United States voters may oppose rail improvements simply because Amtrak is a quasi-governmental agency and rail advocacy may be perceived as contributing to central government encroachment. For many of these voters, any expansion of government is undesirable and must be blocked (Gopnik, 2015).

A new entrant into the passenger rail business (the airlines) would strengthen advocacy for improved passenger rail service, would reduce the fear of central government monopolistic interference in yet another aspect of society, and would add pressure for infrastructure and service improvements in those places where improved passenger movement by rail would make sound economic sense. Amtrak would not be the only voice for rail improvement advocacy but would be joined by a growing number of private sector airline voices. Airlines would also have opportunities to consider a way to save and grow passenger traffic for short distance movements using a potentially less costly passenger movement alternative. The introduction of at least one or two successful examples would highlight the benefit of having a second option for airline passenger growth and revenue enhancement rather than clinging only to short haul plane routes.

Passenger rail deployment controlled by the airlines would also allow the planners to consider ways to significantly improve the passenger experience. Rail service could allow more innovative ways for airlines to serve and attract added passengers. Expedited security screening for passengers who would not be traveling by air is an example. Providing ways for passengers to check luggage through to a final destination or to bring all luggage with them as "carry on" would be another. Better seating, better lighting, more legroom, improved washrooms, expanded food choices, on-board business services, alternative recreational or entertainment options, etc. are all examples of ways to enhance the travel experience. The ultimate travel benefit, aside from getting from one place to another is the opportunity for travelers to convert time lost in driving to productive work or improved quality leisure time. In short, this scenario gives the airline industry a chance to redefine and improve significantly on the typical mass transit experience, particularly the "sardine can" experience of most current air travel.

For Amtrak as well, a partnership with airlines similar to what is being crafted by Iowa Pacific, could be a significant win. Amtrak would still be needed to negotiate right of way agreements with freight railroads and to provide power units or provide licensing agreement to others to haul passenger cars over rail corridors. All these services provide revenue opportunity for Amtrak. The prospect of hauling twelve trains per day each way between Chicago and Indianapolis, for example, certainly provides more revenue potential for Amtrak, even taking into consideration the revenue earned by Iowa Pacific, than its current model of very limited service.

If we think of Amtrak as a partner that negotiates arrangements with freight railroads, purchases rights of way for its services, contracts for and licenses the construction and use of rail infrastructure, etc., it seems that there could be a much greater role (and greater revenue opportunity) for Amtrak to play in facilitating rail expansion. Just as the trucking industry could not have grown and expanded to the extent it has without the interstate highway system, so too the growth of airline passenger movement on rail will require governmental (or quasi-governmental) support for infrastructure development and enhancement. An Amtrak role in helping to create an alternative to the interstate highway system could spark the creation of a passenger movement explosion just as potent as trucking industry growth spawned by the interstate highways. With little added effort, this newly developed infrastructure could also be used to leverage and advance the express parcel movement business on many of these same short distance routes, thus further reducing the pressure on the interstate highway system and offering additional Amtrak licensing revenue.

Another benefit for airline-high speed rail collaboration would include the continuity and reliability of transportation service even in bad weather conditions. Given that no travel mode is free of interruptions, trains can run when airport runways are shut down and highway travel is dangerous or impassible. This reliable alternative should be attractive to airlines and would allow them to move passengers with greater schedule certainty.

One added reason that airport-to-airport train service can be attractive is that airports already have interfaces for light rail, bus service, and vehicular parking. If airlines also provided fully integrated rail service access at the two terminal endpoints, the potential attractiveness to passengers should be apparent. In addition, rail passenger service "gates" at an airport, unlike the gates providing service for planes, should be able by design to move far more passengers and baggage in less time than the typical 30 to 45 minute airplane gate turnaround time. So, instead of a typical airline gate processing two (2) planes per hour and approximately 250 passengers, a rail passenger gate should be able to process six (6) train departures per hour and up to 1,200 passengers. These time enhancements would leverage the other benefits of terminal access and free up short haul plane gate slots for more profitable long distance flights.

Frequency and reliability are key factors in attracting passengers, but the Chicago to Indianapolis example allows a stop in Chicago without the need to fight expressway traffic or pay for downtown Chicago parking. It seems that the convenience factors created by the opportunity to include two or three added stops (and especially downtown Chicago) together with security screening and baggage handling at these added locations, provide a level of attractiveness that should generate significant added riders and revenue.

Determining whether the proposed multi-modal airline business model would work requires further analysis. The planning and preparation developed to this point by INDOT, Amtrak, and Iowa Pacific Holdings could serve as a starting point. In fact, to be fully attractive to them the air carriers might require integrated passenger and baggage handling access for rail service at airport terminals as suggested in this article. That change would require support of airport authorities in addition to Amtrak and INDOT and would entail significant capital costs and infrastructure enhancements, particularly if airport security screening, integrated baggage handling, and ticketing were incorporated into at least some of the rail departures.

Ultimately, of course, the attractiveness of this proposal for the airlines is the possibility of increased revenues accompanied by decreased runway and gate congestion. The business case benefits for airlines rest on a comparative analysis of replacing plane costs with train costs, and evaluating what added revenue opportunities could accrue from the new model. In addition, freeing up gate slots by moving short haul passengers from plane gates to train gates at airports allows airlines to increase the number of more profitable long haul routes while still serving the same (or even increased) numbers of short haul passengers on rail.

An additional bonus for a governmental agency such as INDOT involves the possibility of shifting significant quantities of traffic from interstate highways to rail. This represents potential savings on highway construction and maintenance costs. For example, would shifting 100,000 travelers per month from Interstate 65 to other modes eliminate the need to build a third lane? What would be the savings in doing so, and how does that compare to the costs of providing alternative infrastructure? As population grows, the pressure on existing urban road systems and interstate highways will only grow. From a societal benefit perspective, reinvigorating old transportation modes in ways that hold the possibility of reducing traffic congestion and moving more people more efficiently, with improved safety, at lower total cost, and with greater satisfaction through innovative, collaborative models such as airline-high speed rail needs to be thoroughly and seriously explored.

REFERENCES

Bloomberg Business. (2015). *Company overview of Iowa Pacific holdings, LLC.* Retrieved from http://www.bloomberg.com/research/stocks/private/snapshot.asp?privcapId=13500461

Elking, I., & Windle, R. (2014, Fall). Examining differences in short-haul and long-haul markets in US commercial airline passenger demand. *Transportation Journal*, *53*(4), 424–452. doi:10.5325/transportationj.53.4.0424

Global Air Rail Awards. (2012). *Case Studies: Intermodal Partnerships*. Retrieved from https://www.globalairrail.com/images/content/publications/casestudies/Winners_Case_Studies_Partnerships.pdf

Gopnik, A. (2015). *The plot against trains*. Retrieved from http://www.newyorker.com/news/daily-comment/the-plot-against-trains

Greyhound. (2016). Retrieved from https://greyhound.com/Express/Default.aspx

Hong Kong Economic Journal. (2016). *Airlines explore air rail business collaboration*. Retrieved from http://www.ejinsight.com/20140113-airlines-explore-air-rail-business-collaboration-paper-says/

Hu, M., Patel, A., Simmons, J., & Tsu, J. (2014, April 29). Feasibility of the implementation of a high-speed rail system – Phasing out regional air travel. *AOSS 480 – Final Report.*

Indiana Department of Transportation. (2015). Average daily traffic and commercial vehicles interactive map. Retrieved from https://entapps.indot.in.gov/TrafficCounts/

Indianapolis International Airport. (2016). Retrieved from http://ind.fltmaps.com/en

AASHTO Journal. (2015). *Update: INDOT extends 'Hoosier State' passenger train support through April.* Retrieved from http://www.aashtojournal.org/Pages/031315Hoosier.aspx)

Massachusetts Institute of Technology. (2015). *Global airline industry program: Airline industry overview*. Retrieved from http://web.mit.edu/airlines/analysis/analysis_airline_industry.html

MegaBus. (2016). Retrieved from http://us.megabus.com/Default.aspx

Milmo, D. (2012). Virgin empire: from selling records to running trains and airline. *The Guardian*. Retrieved from http://www.theguardian.com/business/2012/aug/15/virgin-train-richard-branson

Murphy, D., & Meilus, A. (2012, January). Measurement of normalized change in demand for short-haul O&D commercial air rravel from 1995 to 2010. *Paper presented at Transportation Research Board 91st annual meeting*, Washington, DC.

Nixon, R. (2012). *Frustrations of air travel push passengers to Amtrak*. Retrieved from http://www.nytimes.com/2012/08/16/business/hassles-of-air-travel-push-passengers-to-amtrak.html?_r=0

Yglesias, M. (2015). *Why American trains are so bad*. Retrieved from http://www.vox.com/2015/6/12/8764819/why-american-trains-are-bad

Chapter 8
Blended Approach:
Efficient Track Structure for High Speed Rail

Richard Harnish
Midwest High Speed Rail Association, USA

F.K. Plous
Corridor Capital, USA

ABSTRACT

High Speed Rail development is an increasingly significant, interesting topic in the present and expected to continue to grow even more in the future. Implementation of high-speed rail would require the right type of track network system that would need to be maintained and/or built, as necessary. In this chapter, the author recommends that the United States will need to use the Blended or Building Block Approach to develop high-speed rail. He presents several examples of blended infrastructure worldwide for high-speed rail, which would offer a range of tools here in the United States for high-speed rail development.

INTRODUCTION

High-speed rail has been late to develop in North America, and one of the reasons is that it is widely misunderstood. The majority of key decision makers, along with the media that influence them and their constituents, believe that high-speed-rail systems are "too expensive" and take "too long to build." They assume that bil-

DOI: 10.4018/978-1-5225-0102-2.ch008

lions of dollars must be spent over a period of as long as 10 years before a new high-speed rail line/track system network is open over its full length of 200-400 miles. A common misconception is that high-speed rail is a stand-alone system, separate from the conventional network. But that scenario actually is rare. In fact, only Taiwan's system is completely separate. All other systems use the "blended approach" in which trains use a combination of high-speed and conventional tracks in a single journey. It is similar to cars and buses using both Interstate Highways and local roads in a single trip. Most of today's high-speed rail systems were not built end-to-end as "pure," stand-alone, high-speed rail alignments that stood empty until the last mile of line was completed.

In fact, of the nine nations that have built high-speed rail lines, only two have built stand-alone systems that were not incorporated into the existing national railroad network. All of the others were constructed as "blended systems" in which the new high-speed segments were constructed as segments, or "building blocks." These are connected to one or more parts of the existing track alignment so as to get service started sooner and give the traveling public a taste of high-speed-rail travel over a part of the route before the full route was completed.

This segmental approach mirrors what happened when the U.S. began constructing the Interstate highway system in 1956. Because the entire 42,500-mile superhighway network was scheduled for completion over a 20-year period, the Interstates were opened in segments, with the earliest (and easiest) of them built in rural areas, and the difficult and more expensive urban stretches coming only at the end of the project. As each 5-, 7- or 10-mile rural segment was completed, state highway departments erected signs diverting motorists off the older federal and state highways so they could access the new Interstate, incorporate the latest segment it into their trip and test the new concept of safe, high-speed driving on a grade-separated, limited-access superhighway with no stop lights. On reaching the end of the completed segment, motorists were directed back onto the old highway.

The result was that even the most skeptical taxpayers found they really enjoyed and appreciated the new Interstates and began pressing Congress for funding to finish the entire network as soon as possible. The Midwest High Speed Rail Association believes that a similar "building-block" approach should be used in promoting and constructing an Upper Midwestern high-speed rail network with its hub at Chicago (Midwest High Speed Rail Association, 2016).

In some cases, existing railroad alignments have spare right of way on which high-speed track can be constructed now alongside existing conventional track used by freight and Amtrak trains.

Other segments are occupied by active railroad track that can be upgraded for joint freight and passenger-train use with the passenger trains traveling at 80, 90 or 110 miles per hour. Some locations offer abandoned railroad right of way that can be brought back into service as high-speed track.

And in most big cities, high-speed trains will travel over conventional track at conventional speeds while traversing densely settled urban neighborhoods en route to a large downtown commuter-rail or Amtrak station. It is not necessary to build a completely new high-speed railroad line in order to begin operating a successful high-speed rail service.

The Midwest High Speed Rail Association shows advocates and decision makers how some of the key high-speed rail lines worldwide were constructed in building blocks and opened incrementally in order to minimize public expenditures while launching high-speed-rail sooner and accelerating public support for completion of the project. Everywhere this approach has been followed, it has succeeded. The opening of high-speed rail service over a small part of the route attracted more passengers immediately and generated popular pressure on public officials to complete the remaining and more difficult segments as fast as possible. The Association believes that this inventory can serve as a guide to how an Upper Midwestern high-speed-rail system with its hub at Chicago might be assembled and placed into service on a piecemeal basis over a 10-year period (Midwest High Speed Rail Association, 2016).

Each nation's railroad has taken a different approach to developing its high-speed networks in response to diverse needs. The common thread is that by implementing individual projects as a series of building blocks towards a long-term plan, immediate improvements can be enjoyed while working towards a very powerful system. This author contends that for high-speed rail to develop in the United States, a comprehensive blended approach will be needed, with initial focus placed on a few key segments. The following examples offer various approaches to draw on.

Approaches Used in Different Countries

Japan: Extending Service beyond the High-Speed Line

To Japan goes the honor of opening the first true high-speed rail line, the *Shin Kan Sen*, or New Trunk Line, connecting Tokyo with Osaka in 1964. Along with the high-speed railroad opened in Taiwan in 2007, it is the only high-speed rail line that is physically separated at all points from the nation's existing railroad system.

A total stand-alone high-speed railroad was necessary in Japan because the New Trunk Line was designed for standard-gauge rails spaced 4 feet 8 1/2 inches apart, while the existing Japanese railway system used a narrow gauge on which the new high-speed trains would not fit. The only sense in which the *Shin Kan Sen* trains can be said to "use" the existing railway system is at certain large metropolitan stations where for space reasons the high-speed tracks had to be fitted onto a bridge structure directly over the earlier narrow-gauge tracks rather than being given their own footprint alongside the earlier tracks. Over the next 50 years, this totally discrete high-speed system has been extended over newly created alignments to a total length of 1,625 miles embracing three of Japan's four home islands.

But Japan's high-speed rail system no longer is totally isolated from its earlier narrow-gauge rail alignments. The complete break with the conventional network allowed the railroad to create a very productive and reliable network that now covers 1625 miles, but the network is no longer completely separate. Two branches, the Akita *Shin Kan Sen* and Yamagata *Shin Kan Sen* were created by upgrading and converting narrow-gauge lines to standard gauge. These two branches, the Akita *Shin Kan Sen* and the Yamagata *Shin Kan Sen* have a standard-gauge railway overlaid on the existing narrow-gauge line. When eastbound *Shin Kan Sen* expresses reach these junctions, the forward section of the train is separated from a differently engineered trainset coupled to its hind end. The forward section continues to other destinations on the standard-gauge high-speed route, while the rear section continues on the narrow-gauge branch at a speed of 80 mph. Because the loading gauge—the distance between the track and the platform—could not be changed as long as narrow-gauge local trains continued to operate on these branches, the *Shin Kan Sen* trains operating over the branches use a narrower car body with an automatic bridge plate at each door that spans the gap between the car and the platform.

These two branches are served by 6 or 7-car trains that operate in tandem with a mainline train between Tokyo and the junctions of Fukushima and Morioka at up to 200-mph. There the two trainsets are separated with one trainset continuing eastbound on the mainline and the other proceeding down the branch. The trains are restricted to 80-mph on the branches and share track with commuter trains and freight trains. While the track gauge was widened, the loading gauge was not. As a result, the trains that operate on these branches are not as wide as other bullet trains. An automatic bridge plate fills the gap between the train and the platform at high-speed platforms. The JR East is testing gauge changing trains to allow high-speed trains to serve more cities on narrow change lines without the expense of rebuilding those lines (Figure 1).

France: Adding Key Segments to a Robust Conventional Network

SNCF, the French National Railroad, did not face the challenge of differing track gauges since its network was entirely standard gauge. Also, most lines were already electrified. Therefore, SNCF could add key segments of high-speed track into an already existing network.

Figure 1.

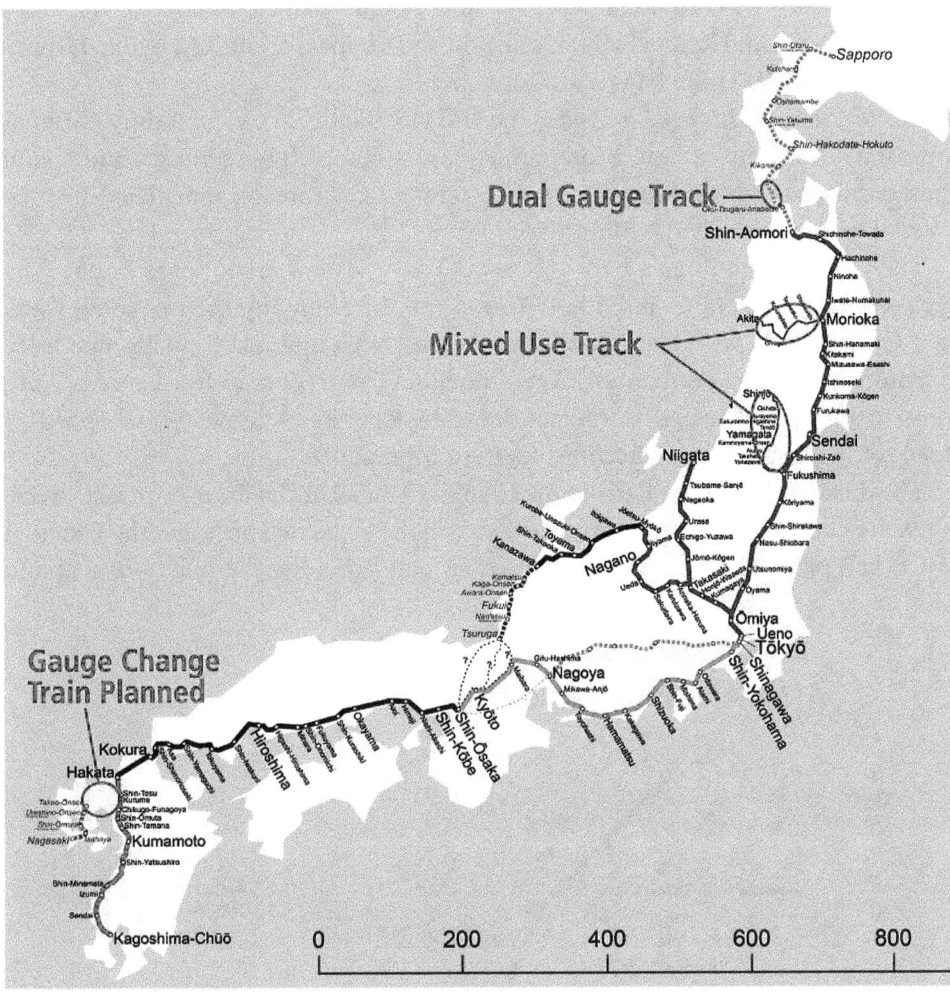

The first high speed line, the LGV Sud Est (Lignes à Grande Vitesse or high-speed line), did not actually reach Paris or Lyon and its impact stretched to the entirety of southern France and beyond. The TGV (Train à Grande Vitesse or high-speed train) not only achieved 168-mph on the LGV, but its low-slung, lightweight, and articulated design created a huge performance increase on conventional track. The TGV is more stable on rough track, can take turns at higher speeds, gets into and out of stations faster, and costs less to operate than the conventional trains it replaced. As a result, the TGV could use existing tracks to reach the historic Gare de Lyon in Paris and Lyon's two main stations. TGV's also used existing or classic lines to reach far beyond the initial line. Soon after start up, the TGV was serving many cities throughout the south of France with several daily trains as far as Geneva and Rome. The LGV was extended twice and now extends to the outskirts of Marseilles with a bypass around Lyon via Lyon's airport and a bypass around Avignon. Express trains cover the 500 mile trip in just three hours.

The second high-speed service, the TGV Atlantic, took a slightly different approach. The Paris station, Montparnasse, was completely reconstructed and an unused railroad right-of-way was used to get the LGV directly to it. The LGV was covered with a linear park within Paris.

The LGV was built as a bypass around Tours. A new station was built on the bypass and shuttle trains provide the final link into the city. Just beyond Tours, the LGV connects with the classic line that was been upgraded to 125-mph as far as Bordeaux. A new segment of LGV stretching from Tours to Bordeaux is under construction now. Beyond Bordeaux the trains operate at conventional speeds to Hendaye, with some trains heading south to Montpellier.

The third LGV, linking Paris to the Channel Tunnel at Calais, used classic tracks to reach Gare du Nord but new tracks between the Paris outskirts and the Chunnel. SNCF had planned on building a bypass around Lille, but the mayor was able to

Figure 2.

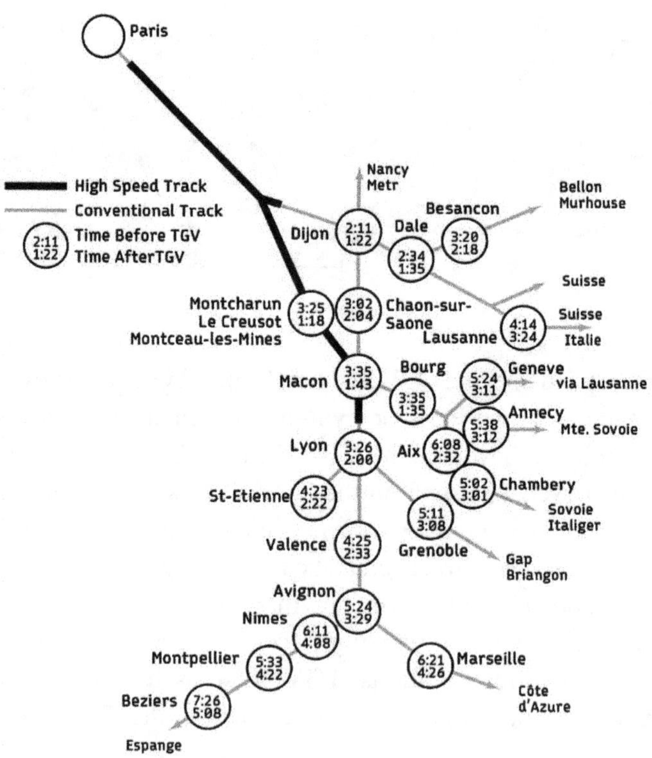

exert political pressure to have the new line serve the city center. As a result, the new high-speed station is just blocks from the old station. Belgium constructed a connecting high-speed line from Lille to Brussels. Thalys branded TGV trains extended the trip from Paris through Brussels to Amsterdam. Belgium and the Netherlands have added segments of high-speed line to reach Amsterdam.

The LGV Est follows a similar strategy to the LGV Sud, originating from Gare l'Est and trains fanning out from the LGV, traveling as far as Frankfurt, Germany. A second segment to reach Strasbourg is being tested at this time (Figure 2).

The new LGV, however, was not a continuous 186-mph railroad from end to end. The new high-speed trains used SNCF's existing conventional tracks from the historic Gare de Lyon in downtown Paris to a point in the suburbs where they switched to the new high-speed line. At Lyon, they do the same thing, switching from the new high-speed line to the older route into the two main stations. The slower speeds on the older stretches did not materially impede train performance. In France, remember "conventional" usually means a passenger-train speed limit of 125 mph. Today, 21 daily TGVs in each direction cover the 284-mile route between Paris and Lyon in a flat two hours. Incidentally, three Midwestern rail corridors—Chicago-St. Louis, Chicago-Detroit and St. Louis-Kansas City—also measure 284 miles from end, but the current running times are about 5 1/2 hours, and after current upgrades are completed on the Chicago-Detroit and Chicago-St. Louis corridors travel times are expected to be only about an hour shorter than they are now.

The original TGV line of 1981 proved extraordinarily successful and now carries the bulk of all travel between Paris and Lyon. Air travel no longer is offered in this corridor because the TGVs beat the plane. Based on the success of the first *Ligne a grande vitesse* of 1981, new high-speed lines have been opened and earlier ones extended, while the TGV trains themselves often are routed for considerable distances over connecting conventional main lines that limit the TGVs to less-than-TGV speeds. The original Paris-Lyon TGV line has been extended to just outside Marseille, and many of the TGV trains that use it continue on over non-TGV track to the historic Riviera communities of Toulon, Cannes and Nice. Because of their light weight and articulated suspension, the TGV trains are able to negotiate the tight curves on conventional track with a high degree of comfort, so they remain attractive to passengers even when they cannot deliver a 186-mph trip over an entire itinerary (Figure 3).

This blending of high-speed track with conventional track to offer service in TGV equipment over an entire route has since spread to the northwest and southwest quadrants of France, as well as east and north to offer high-speed service into neighboring countries such as Britain, the Netherlands, Belgium and Germany. In January 2016 workmen finished installing concrete ties and welded rail on the

Figure 3.

TGV Atlantique extension from Tours to Bordeaux, and another crew of workmen is now following behind erecting the catenary. Another new line east from Paris to Strasbourg is now open to test trains but not to the public.

The flexibility of the blended approach becomes apparent with the LGV Rhin-Rhone which seems not to connect any major cities, but as part of the overall network is a valuable link between eastern France/western Germany and the Mediterranean.

England: High-Speed Commuter Trains

The blended approach played a critical role in connecting England to France. England was not committed to high-speed rail but was willing to participate in building the Channel Tunnel. The speeds gained on the French side were enough to make the train competitive in the Paris - London market even though they would travel a low speeds on the English side.

The TGV design was modified in three ways to serve the English market. It was narrowed for the smaller loading gauge of the English network, and it was made so that all the passengers could move into one half that could be easily separated in case of a train fire in the tunnel, and third rail shoes were added to draw from the 600-volt third rail in England.

Blended service played another important role when the English built HS1, a new high-speed line directly into a completely reconstructed St. Pancras station. The line was built to serve 140-mph commuter trains that depart from a separate section of St. Pancras, use the high-speed tracks to leave London, and then divert from the high-speed line into third-rail territory to serve the many suburbs in Kent (Figure 4).

Figure 4.

High Speed Rail in the Low Countries: A Piecemeal Effort

Well before the long-awaited Eurotunnel under the English Channel made direct London-Paris high-speed train service available in 1994, plans were afoot to extend service into Belgium and the Netherlands via the new TGV station erected in the northern French town of Lille, very close to the Belgian border. An entirely new *Ligne a grande vitesse* known as *LGV Nord* was built from just north of Paris to Lille near the Belgian border, then northwest to Calais where the line entered the Eurotunnel. Just south of Lille, a major junction was built and high-speed track extended across the Belgian border, which lies only about 15 miles northwest of Lille.

The new high-speed line does not extend all the way into Brussels, however. It ends at Halle, about 12 miles south of the capital, where high-speed trains switch to an adjacent conventional track for the rest of the trip. Even when the last segment is completed, however, it will not reach all the way to Brussels Midi Station. Instead, the high-speed trains will switch onto the existing track network about a mile and a half south of the big downtown station. Through trains continuing on to Amsterdam will cross an even bigger gap north of Brussels, running on about 25 miles of conventional track before entering another section of high-speed track about 2 miles south of downtown Antwerp. This high-speed section extends north of Antwerp toward Amsterdam for another 45 miles, where it breaks for about 3 miles just short of Rotterdam before resuming again in Rotterdam and running 18 miles to the great air-rail interchange station at Amsterdam's Schipol Airport. The last six miles from the airport to downtown Amsterdam are run on conventional track. Even though high speeds cannot be attained from end to end over the full route from Paris to Amsterdam, high-speed trains run through, and passengers understand themselves to be making a high-speed train trip between the two cities. Because the *LGV Nord* line makes a stop directly under the passenger terminal at Paris's Roissy-Charles de Gaulle Airport, the TGV trains have totally replaced Brussels-Paris air service since 2000. High-speed track construction has been going on in the Low Countries since 1996 and will not be finished for many years, but travelers understand the long-term nature of infrastructure planning and funding and have made the existing, steadily improving, train service very popular.

Spain – Gauge Changing Trains

As in Japan, Spain's classic network is a different gauge than its high-speed track. In this case, its older network is wider gauge than international standard gauge. It was decided to build the new high-speed lines in standard gauge in order to allow connectivity with France.

Like Japan, new tracks had to be constructed into mainline stations, which were greatly expanded to meet the traffic demands created by high-speed rail. Unlike Japanese Railways, Renfe, Spain's national railroad, decided to extended high-speed service beyond the high-speed line right away. They did this with gauge changing trains made by Talgo. Automatic gauge changing machine were installed just east of the Córdoba and Seville stations to allow trains to continue on the old tracks to Malaga, Cadiz, and Huelva. Like the TGV, the Talgo delivered increased performance on the very curvy track of Spain.

The gauge changing trains were initially limited to 125-mph, while the trains remaining on high-speed track were allowed to reach 186-mph. Those limits have been raised to 155-mph and 205-mph. Renfe created two brands to distinguish the services: AVE for the fastest trains and ALVIA for the gauge changing trains. A third brand, AVANT, denotes 155-mph commuter trains that use the high-speed line.

Renfe added an additional twist to reach the far western and northern cities along the Atlantic Coast whose tracks have not yet been electrified. By an additional unit holding a diesel generator to their gauge-changing trains, Talgo was able to create a trainset capable of 155-mph on high-speed track, better performance on conventional track and 90-mph speeds on non-electrified track (Figures 5 and 6).

Germany: A Complex Network

A very different political geography led to a very different approach to high-speed rail in Germany. Trains in Germany operate in a complex network, using a mix of high-speed, upgraded, and conventional track in a single journey. The Japanese, French, and Spanish networks are largely radial with core lines extending from a large political and economic center, which are key segments of high-speed line that funnel traffic from a broad area.

Figure 5.

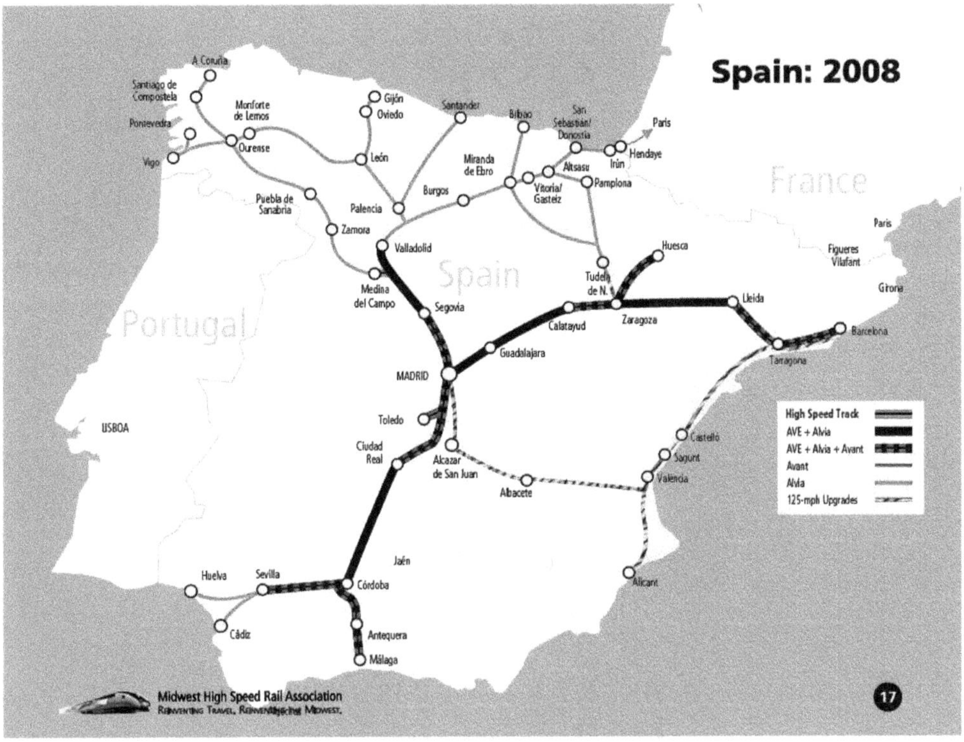

Figure 6.

Germany does not have a single core city and instead has five primary centers spread across the country. The resulting railroad forms more of a grid network rather than the hub-and-spoke networks in France and Spain. Only the 186-mph Franfurt - Cologne line is dedicated to high-speed trains, and it serves as a funnel for trains coming from Brussels, Amsterdam, and other locations. The trains enter both Cologne and Frankfurt on shared tracks. Other segments, like to Hamburg - Berlin line,

Figure 7.

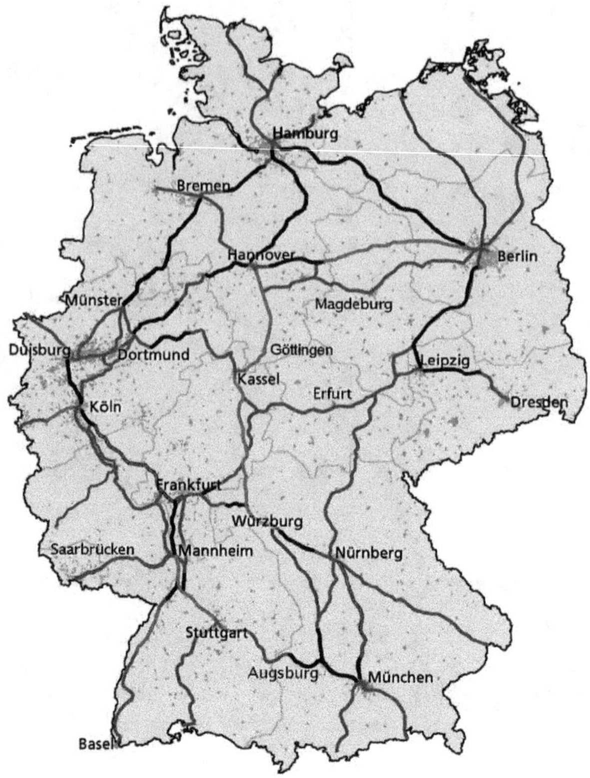

were built for a maximum speed of 155-mph so that they can be used for freight. Slower commuter trains share the 205-mph Munich-Nuremberg line. Frequency of service and well-coordinated connections make up for the lower speeds (Figure 7).

Korea Builds Incrementally

South Korea's mountainous territory virtually dictated that a complete high-speed rail alignment could not be built in one continuous project. High-speed rail lines are supposed to be straight—or nearly so—but building a straight line through mountains requires a multi-year effort. And building across the valleys between the mountains means transporting millions of cubic yards of fill—usually made up of the spoil excavated during the tunnel-building process—to carry the tracks on a high embankment over the lower terrain. Also, the solid fill alone was not enough. Thirty-five percent of the mileage had to run on high viaducts over the rice paddies. When the Republic of Korea began planning its first high-speed rail line in the 1980s it was understood from the start that it would have to be opened in segments.

The route selected was to connect the capital, Seoul, in the northwest corner of the country just south of the North Korean border, with Busan, the nation's second-largest city and its main seaport on the southeast corner of the 180-mile wide peninsula. The distance between the two cities is about 275 miles, roughly the same as the rail corridors connecting Chicago with Detroit and St. Louis. In 1985, the diesel-powered trains connecting the two cities took 4 hours and 10 minutes to cover the distance, but even at that pokey speed they were carrying 57 million passengers per year. South Korea aimed to cut this time to about two hours.

Construction began in 1992 with a 35-mile segment between Chonan and Daejun, just north of the mid-point of the route. At the same time, the Korean rail company bought two TGV train sets from the French that it could test on the completed line segment, which was supposed to open in 1999. But construction did not go smoothly, and after protests from historic preservationists the railroad was forced to re-route the line around Kyongju, ancient capital of the Silla kingdom. Progress was further interrupted when the construction firm hired to build the new line turned out to lack the necessary skills and had to be replaced with a new contractor.

In the late 90s the Korean economy slid into a financial panic that led to a change of government and postponement/re-evaluation of many government-sponsored projects, including the buildout of the Seoul-Busan high-speed line. The purchase of TGV trains from France continued, but development of true high-speed rail infrastructure was cut back to the Seoul-Daegu segment. However, even though a new TGV alignment was not developed between Daegu and Busan at the time, the existing alignment was electrified so that TGV trains—if not total TGV service-could be offered over the entire route. On March 4, 2004, TGV service opened over the whole route, cutting travel times from 4 hours and 30 minutes to 2 hours and ten minutes.

The surge in ridership that followed made it clear that the final high-speed segment had to be constructed. Today the full TGV is open from Seoul to Busan and carrying virtually all travel between these two points. Politics, finance, and historical-preservation issues plagued the project from end to end but did not stop it, and opening the line in segments only whetted the public's appetite for completion.

Information Applicable for the United States

Pursuing blended service will allow the US to gain huge benefits from high-priority segments of high-speed line before the complete network is built and those segments can gain greater volume sooner by serving a broader network. California is leading the way with the CalTrain electrification project and has a big opportunity in the Central Valley. Other opportunities exist in the Midwest. Two challenges will need to be overcome to make it work.

High-Performance Trainsets

A new train, capable of both providing high-performance on rough freight track and operating at very high speed on dedicated tracks, will need to be developed. Outdated safety regulations and design practices in the United States have resulted in trains that are very heavy causing high operating and track maintenance costs. The Federal Railroad Administration has begun developing new regulations, but much work remains to make a train that is more efficient and safer than existing designs.

A New Relationship between Host Railroads and Passenger Train Operators

The current relationship between Amtrak and the host railroads is broken and has created neither commercially nor politically viable service offerings. The core of the relationship is the "incremental cost" formula that allows Amtrak to demand the highest levels of service while paying the lowest rates. A new relationship that creates incentives for the railroads to invest in faster, more frequent and more dependable trains will need to be developed.

Box 1. California map

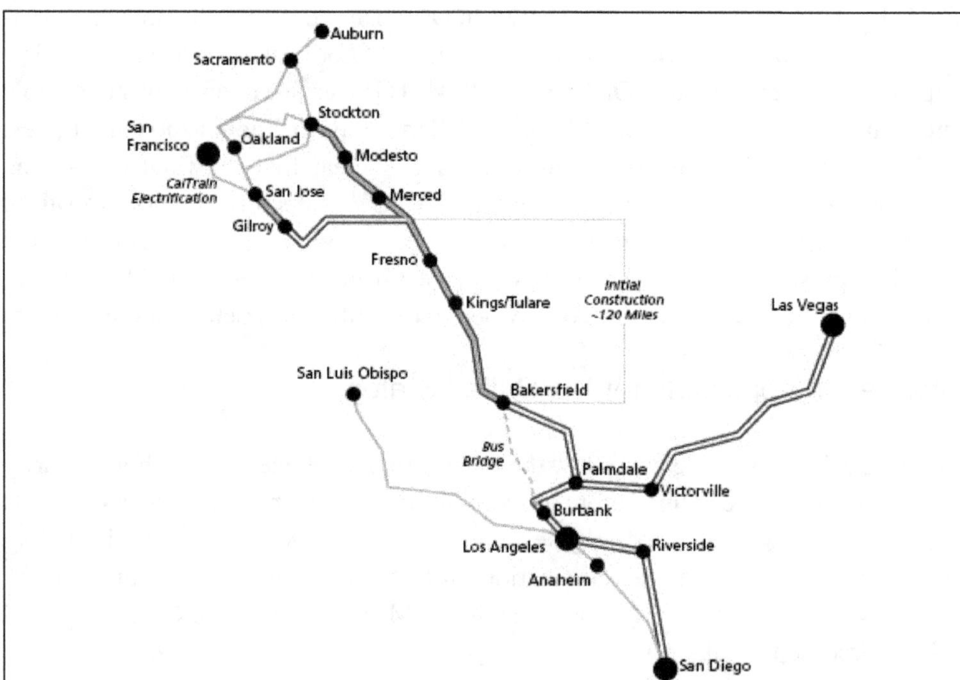

Opportunities in the United States

The term "Blended Approach" was first coined in California to help explain its phased implementation plans. The California High Speed Rail Authority has begun constructing the middle segment of a Los Angeles – San Francisco high-speed network between Modesto and Bakersfield in the Central Valley. The network will use blended service at both ends.

CalTrain

At the northern end, the San Jose – San Francisco CalTrain commuter line is being reconstructed to be shared by commuter trains and high-speed trains. The track will be reconstructed and electrified. New passing sidings will be added. Heavy, diesel powered commuter trains will be replaced with lightweight, electrified trains based on European designs. Platforms will be raised to 50 inches to create level boarding.

Northern Unified Service

Today, there is a loosely connected network of Amtrak and commuter trains linking Oakland, Sacramento, Stockton to Modesto and other cities. With a new, blended-service trainset, the Modesto – Bakersfield segment of high-speed line could be a catalyst for upgrading these services with faster and more frequent schedules.

Figure 8.

The Midwest Hub

The economic geography of the Midwest is very similar to France and Spain with Chicago serving as the hub of a vibrant, nine-state region. As such, it a great location for a hub and spoke blended network built from the center out. Here is one way the network could progress.

CrossRail Chicago

The core of the system could be the Metra-owned Electric route. The Electric is a very unique railroad in North America. The mainline and branches are dedicated to passenger trains, electrified and feature 50 inch platforms for level boarding. The mainline stretches 30 miles to the edge of the urban area without a single level highway crossing and is expandable. If modernized, the Electric could be the access point for high-speed routes from the east and south.

Other mostly Metra-owned segments could be reconstructed to similar standards to create a multi-purpose, high-density trunkline dedicated to passenger trains linking the Electric to Chicago Union Station and O'Hare airport (Figure 8).

The Midwestern T-Bone

The end of the Metra Electric at University Park could be the start of the first high-speed line in the Midwest, built in several segments linking Chicago to the University of Illinois at Urbana/Champaign, the state capital in Springfield and then St. Louis. This would be roughly the same distance as Paris – Lyon, Madrid – Seville and Tokyo – Osaka. A branch from Champaign to Indianapolis would form the shape of a T-bone steak (Figure 9).

Upgraded Segments

There are several opportunities to extend this core network by partnering with existing railroads to upgrade and expand existing lines. For example, the lines from Indianapolis to Cincinnati and Louisville are relatively straight. It may be possible to work with those railroads and the local communities to add second main tracks and separate highway crossings to allow for 125 mph speeds. Other segments such as Champaign – Carbondale that use high-density freight lines could be expanded to allow additional frequencies at conventional speeds.

Blended Approach

Figure 9.

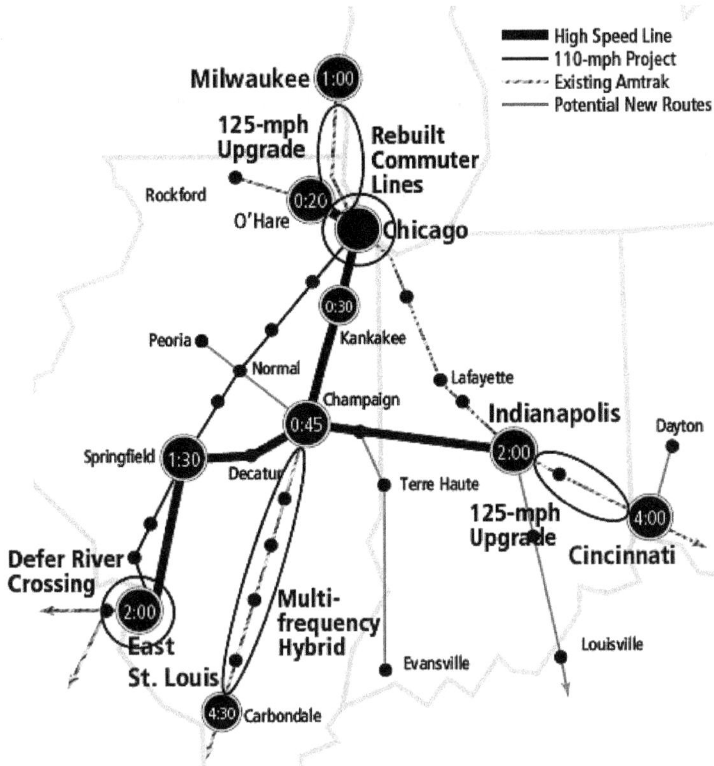

In conclusion, the author recommends that the United States, similar to several other countries in the world, would optimally, efficiently, and expediently enhance high-speed rail development through a blended approach for its track structure system. Further, it would be the most cost-effective method to implementing high-speed rail by using existing track structures rather than developing new lines.

REFERENCES

European Trains. (n. d.). *Rail Europe.* Retrieved from https://www.raileurope.com/european-trains/

High-speed rail in Germany. (n. d.). Wikipedia. Retrieved from https://en.wikipedia.org/wiki/High-speed_rail_in_Germany

Hughes, M. (2015). *The Second Age of Rail: A History of High-Speed Trains.* London, UK: The History Press.

Midwest High Speed Rail Association. (n. d.). Retrieved from http://www.midwesthsr.org/

Shinkansen Map. (2016, February 8). *Wikimedia*. Retrieved from http://commons.wikimedia.org/wiki/File:Shinkansen_map_201503_en.png

Southeastern (train operating company). (2016, February 8). *Wikipedia*. Retrieved from https://en.wikipedia.org/wiki/Southeastern_(train_operating_company)

Chapter 9
High Speed Rail in Canada:
Feasibility Studies, Influential Factors and International Comparisons

Jaydeep Balakrishnan
Haskayne School of Business, University of Calgary, Canada

Tianyuan Zhu
Haskayne School of Business, University of Calgary, Canada

Peter Wallis
Van Horne Institute, Canada

ABSTRACT

High Speed Rail (HSR) is gaining popularity around the world for a variety of reasons. However, North America lags behind and in particular Canada, where there is no HSR. There has been discussion of implementing HSR in Canada since the 1970s in two possible corridors (one in central Canada and the other in Alberta). In this chapter, the feasibility studies that have been done on both corridors are discussed. In this respect, the focus is on the recent studies, from different sources including the Alberta Legislature, Ontario and Quebec Governments, consulting companies, and a non-profit transportation institute. The factors that influence the viability of HSR are discussed and reference is made to HSR implementations in other countries that could inform the development of HSR in Canada. On a positive note, it appears that the Ontario government plans to move ahead with HSR in part of the central Canadian corridor.

DOI: 10.4018/978-1-5225-0102-2.ch009

INTRODUCTION

Passenger rail is an important transport mode in many countries. Rail transportation is considered more efficient in terms of energy consumption and more environmentally friendly, compared to road and air transportation. Further as discussed later, it helps reduce traffic congestion and other externalities, such as pollution and accidents that arise from excessive road travel and transport. (Van Horne Institute, 2004, 2013).

In North America, the railroad network is extensive and transcontinental. However, most of the network is used for freight transport; passenger railway ridership, on the other hand, is lower than in Europe and Asia. For example, as shown in Figure 1, while USA and Canada accounted for 30% of world freight traffic in 2011, the corresponding passenger percentage was 0.33% (International Union of Railways, 2011).

The growth of high speed rail (HSR) has been slower in North America compared to Europe and Asia. In Canada, only two feasible scenarios are seen for the foreseeable future. The first is in the southern Ontario and southeastern Quebec corridor and the second is in the Calgary Edmonton corridor in Alberta.

HSR was introduced in Japan about fifty years ago and has made great strides in recent years throughout Europe and Asia. HSR provides fast, efficient, and hassle-free service that can compete with air transportation over short to medium distances, especially given the additional inconveniences that air passengers have endured post September 11, 2001. It has also been shown to provide economic benefits to regions served (Bernard, Moxnes & Saito, 2014).

Figure 1. 2011 world traffic (International Union of Railways, 2011)

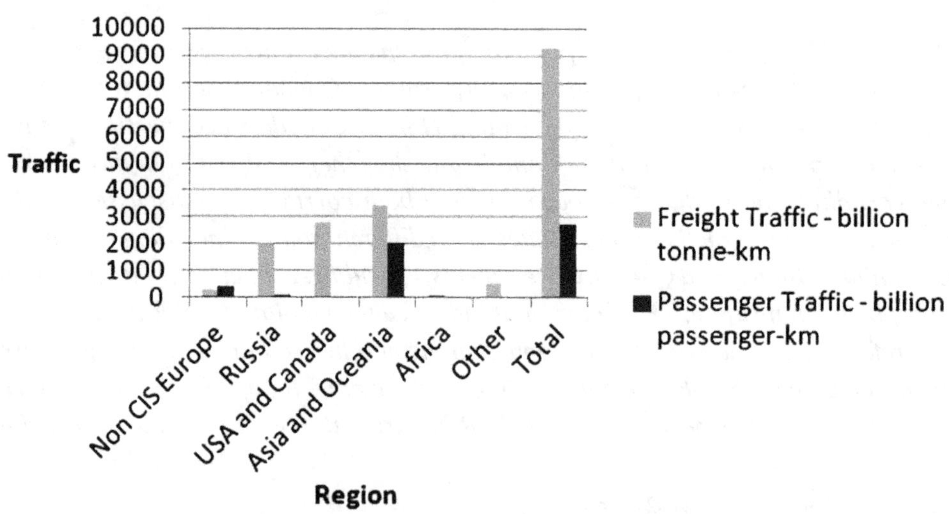

BACKGROUND

The history of the Canadian railway dates back over 180 years, when the first railway in Canada, the Champlain and Saint Lawrence Railway opened outside of Montreal in 1836. Transportation, particularly railways, was an integral aspect of the development of industry in Canada. The laying of a railway line contributed to a place becoming a preferred manufacturing location as it provided that hub cheaper access to larger geographical markets as was the case for the Eastern Townships of Quebec. Prior to railways, the lack of good roads meant that the economy was necessarily local since transporting inbound and outbound material long distances was infeasible. The railways also contributed to manufacturing growth by operating large manufacturing and repair shops (Craven and Traves, 1983).

Before confederation in 1867, railway construction proceeded slowly in Canada. By 1850, the United States had 9000 miles (14500 km) of track while Canada had only 66 miles (100 km) (Pomfret, 1981). This low track mileage was due to the availability of waterways and a lack of government subsidies. After confederation, the government's desire to unify the nation resulted in subsidies for extensive rail construction and contributed to the economic development of the country (Pomfret, 1981). For example, concerned about the possibility of present day British Columbia (B.C.) being annexed by the U.S., one of the enticements offered to British Columbia to join with Canada was the promise of a railway connection within 12 years. Although British Columbia became part of Canada in 1871, it was 1880 before the government formed the Canadian Pacific Railway (CPR) and provided it with land and subsidies to extend the railroad all the way to the Pacific coast. (Balakrishnan, Eliasson & Sweet, 2007)

After nearly two hundred years of development, Canada has 31000 miles (49,400 km) of track in total today, of which only 81 miles (129 km) is electrified. The majority of Canada's railway system uses standard gauge track which is 56.3 inches (1,435 mm), with the exception of the broad-gauge streetcar, the subway lines in different cities and the small lines isolated from the main North American rail network being used in resource industries such as mining or forestry.

The current large and well-developed railway system in Canada is now mainly used for freight transportation. The Canadian National (CN) and CPR are the two major publicly traded transcontinental freight railway systems.

Canada's nation-wide passenger services are provided by a Crown Corporation named VIA Rail. Currently, there are three Canadian cities having commuter train services, to serve the satellite cities: the Montreal area by Agence métropolitaine de transport, the Toronto area by GO Transit, and the Vancouver area by West Coast Express. The public transport within these cities and several others such as Calgary and Edmonton are served by light rail or metro systems. There are smaller railways running passenger trains to remote rural areas such as the Ontario Northland Railway and Algoma Central Railway. In addition, sightseeing trains are run by the Rocky Mountaineer and Royal Canadian Pacific providing luxury train travel for viewing scenery in the Canadian Rockies as well as other mountainous areas in Alberta and British Columbia. In terms of regulation and legislation affecting railways in Canada, it is only the Federal Government (Transport Canada) that is involved. Provinces and municipalities do not have any role in this aspect (Transport Canada, 2014).

THE DEVELOPMENT OF HSR WORLDWIDE AND IN CANADA

Globally, with the rapid development of road and air transport, modal competition is becoming increasingly fierce. In the last century, there have been technological improvements in aircraft and road vehicle designs and their associated transportation systems that have facilitated the growth in air and road passenger and freight transport. HSR is an example of technological advancement in rail transport that has made this mode more attractive.

It is not surprising that in densely populated regions in Asia and Europe, HSR is an appealing transportation option. Further, compared to North America, the rate of car ownership is lower (such as in Asia) or even if it is higher than Asia (such as in Europe), the costs of driving are very high. In addition, public modes of transport are strong in these areas (Hurley and Jones, 1990). Japan, France, and Germany pioneered the HSR revolution. These areas served are highly urbanized with population distribution ideally suitable for rail operations, with each country having railway networks that have existed since the nineteenth century. These three countries developed their own HSR independently and followed different paths with their own technology. Specifically, since Japan introduced its famous Shinkansen system connecting Tokyo and Osaka in 1964, it has experienced a dramatic growth in high-speed rail construction and operation during these years. The Shinkansen system in Japan consists of 1537 miles (2,459 km) of lines with maximum operating speeds of 150-200 mph (240-320 km/h,) carrying 155 million passengers in 2014 (JR Central, 2015). Introducing its TGV (Train à grande Vitesse, High Speed Train) in 1981, France now has four main lines radiating from Paris, containing various branches, 1273 miles (2,037 km) in total, with maximum operating speeds up to 200 mph (320 km/h). These lines carried 114.45 million passengers in 2011 (Pepy & Perren, 2006). Deployed in 1991, the InterCity Express (ICE) in Germany now operates a network of 4291 miles (6,865 km) with speeds up to 188 mph (300 km/h). In 2006, ICE trains transported 67 million people (Albalate & Bel, 2012).

Through the use of technologies developed by the three-pioneer countries or using these technologies as a model for their own developments, numerous other countries have introduced HSR operations during the last three decades. These include Italy, Spain, Belgium, U.S.A. (between Washington and Boston), United Kingdom (through the Channel tunnel), South Korea, Taiwan, China, Turkey, and Russia (EcoTrain, 2011). Particularly, in recent years, HSR networks have expanded rapidly in China, which as of the end of 2013 operates more than 6000 miles (10,000 km) of track (World Bank, 2014).

There are also HSR systems being planned in Europe (Norway, Portugal, Poland, and Sweden); in Africa (Algeria and Morocco); in Asia (India, Iran, and Saudi Arabia); and in South America (Argentina and Brazil) (EcoTrain, 2011).

While so many countries are enjoying the benefits of HSR service, Canada has not kept pace with the rest of the world. Canada is the only G8 country without existing HSR infrastructure and the only G20 country without official plans to construct an HSR line in the coming decades (see Table 1 which shows HSR lines in operation and in planning up to the year 2025).

Ironically, there was a period when Canada was poised to become a global leader in developing HSR technology. Canada had "some of the best technology, manufacturing, and research at the National Research Council, and speed records to lead the world in high-speed rail" by the 1960s (Jeanes, 2009). Canadian firms such as Bombardier (railway coaches) and SNC-Lavalin (construction) are well-known internationally for their expertise. Unfortunately, these factors have never been reflected in the implementation of HSR passenger service in Canada. In 1967, the Turbo Train operated by CN (subsequently VIA Rail) between the years 1968 and 1982 achieved a speed of 172 mph (275 km/h). Even though such speed had never been achieved during regular service runs, this was the closest time that Canada ever came to high-speed passenger rail service. The HSR technology and plans were then abandoned in general when the Crown Corporation faced deep cuts from the federal government. Although a number of proposals for HSR by conglomerates have been put forward to parliamentarians, these proposals never went further than the conceptualization phase due to the large capital investment required under

Table 1. High-Speed rail plans compared (kms) (International Union of Railways, 2014)

Region	Europe	China	Japan	Rest of Asia	USA	World
In operation	7351	11132	2664	1445	362	22954
Under construction	2929	7571	779	1275		12754
Planned	10815	3777	179	3543	777	18841
Total (km)	21095	22480	3622	5012	1139	54549

a condition of fiscal restraint, and the logistical difficulties raised by private or semiprivate ownership of passenger rail.

Though no HSR project has been approved to date, Canadian decision makers have considered the idea of introducing HSR systems in the country's busiest transportation corridors since the early 1970s. The most frequently discussed possible areas for HSR in Canada are the Quebec City–Windsor Corridor (QWC) and the Calgary–Edmonton Corridor (CEC) which are Canada's two most densely populated corridors, covering eight of the country's ten largest metropolitan areas and together accounting for up to 60% of the nation's population (see Figure 2). In both corridors, the introduction of HSR is still a hypothetical issue.

Figure 2. Origin and destination of the QWC and CEC corridors
Modified from: Canada. (no date). Canada, North America: Brock University Map, Data & GIS Library. Available: Brock University Map, Data & GIS Library Controlled Access http://www.brocku.ca/maplibrary/maps/outline/North_America/canada.jpg (Accessed July 20, 2014).

Quebec City–Windsor Corridor (QWC)

The population of the QWC area is 19.7 million (2006) and the distance between the Quebec City and Windsor is 750 miles (1200 km, EcoTrain, 2011). Currently, the QWC area has its own passenger rail service operated by VIA Rail. However, because private freight companies mainly own the tracks, the service speeds are limited to conventional speeds (100 mph; 160 km/h).

More than 16 feasibility studies of HSR in the QWC have been conducted since 1973, which has sometimes resulted in business proposals, by private firms. However, none of these proposals was approved. Following the study by the Ontario-Quebec Rapid Train Task Force in 1991, both Bombardier and ABB proposed plans to parliamentarians. Later, the Lynx Consortium suggested building and operating a bullet train using a Public-Private Partnership (P3) model. This proposal by Lynx consortium claimed that the bullet trains would reduce fossil fuel use in the corridor by 20% by 2025 through using renewable energy sources, and was first recommended by the parliamentary committee to proceed to the second phase. However, this was later turned aside in favor of the cheaper but slower, conventional trains. The latest feasibility study of the QWC was by EcoTrain in 2011, which concluded that only the portion of the route between Quebec City and Toronto would generate considerable net economic benefits by introducing HSR (see Table 2, which shows that the WQC as a whole has a negative NPV).

Table 2. Costs and net present value (NPV) of HSR in the QWC[a] (EcoTrain, 2011)

Representative Costs	
Total Cost of Investment	$21.3 billion
Total Cost of Operations & Maintenance	$520.2 million per year
Representative Benefits	
Net Present Value (NPV) of Government Investments	$18.2 billion for wholly public case $9.2 billion for public-private case
NPV of Benefit to Public Safety	$2.5 billion
NPV of Benefit from Reduced Emissions[b]	$96 million
NPV of Entire Project	**-$992 million**[c]

[a]All values are based on 2009 Canadian Dollars and the 188 mph (300 km/h) technological scenario. All Net Present Value figures are based on a 30-year period (2025–2055).

[b]This would include emissions of both greenhouse gas (GHGs: Carbon Dioxide, Methane, Nitrous Oxide) and Criteria Air Contaminants (CACs: Carbon Monoxide, Nitrogen Oxides, Volatile Organic Compounds, Sulfur Dioxide, and Suspended Particles).

[c]Although the study expects a net loss when the entire corridor is considered, the portion from Toronto to Quebec City was found to produce a positive NPV.

According to the most recent feasibility study of the QWC by EcoTrain, the stations served would be as follows: Quebec City (existing downtown Gare du Palais, plus one new suburban station), Trois-Rivières (new station), Montreal (existing downtown Central Station plus two new suburban stations), Ottawa (existing VIA Rail station), Kingston (new station), Toronto (existing downtown Union Station plus two new suburban stations), London (new station), and Windsor (new station) (EcoTrain, 2011). Stations such as Kitchener-Waterloo were dropped because of the short distance to Toronto and the planned extension of the GO commuter line.

There have been some positive developments recently regarding a section of the QWC HSR corridor. The Ontario Government is planning to move forward on HSR between Toronto and London (Smith, 2014). The project, which could cost up to C$ 3 billion, is part of the government's C$ 29 billion infrastructure plan. Design and environmental impact studies are expected start in fall 2014, taking up to four years. Construction may take a further four years to complete. Consulting company First Class Partnerships (FCP) did a pre-feasibility study for the Ontario government and felt that the project had a strong business case. The company noted that there are approximately 20 million trips per year in the corridor and projected this to rise to 30 million trips within 10 years. It was forecast that 20-25% of this traffic would move to rail. It is expected that the HSR would share existing infrastructure on parts of the line and new infrastructure would be built on other parts of the line. The funding would come from provincial gas and sales tax revenues, from the federal government, and borrowing. Thus, no public private partnership is envisaged. It is hoped that success of HSR on this section of the QWC would provide an incentive for HSR expansion on other parts of the corridor.

This will be accompanied by local light rail transit improvements along the line (Canadian Broadcasting Corporation, 2014). The line would connect Toronto Pearson Airport was well as include the construction of railway tracks, stations, bridges, administrative facilities, and the installation of signaling and safety systems (*World Market Intelligence News,* 2014).

Calgary–Edmonton Corridor (CEC)

The two major cities of Edmonton and Calgary have not been connected by the passenger rail service since conventional passenger rail service in CEC was shut down in 1985 (Greenlaw, 2007). Discussion of a HSR link between Edmonton and Calgary dates back to the late 1970s (Van Horne Institute, 2004). Alberta Government sponsored reports in the mid-1980s and in 1995 concluded that the situation was not as of yet conducive for HSR.

The 2004 study, at the request of the Alberta government, conducted by the Van Horne Institute based at the University of Calgary indicated, "sufficient demand exists today". The objective of the study is to design an approximately 185 mile (300 km) rail line to serve both downtowns and Red Deer. More recently, an economic assessment by Transportation and Economics & Management Systems, Inc. (TEMS) in 2009 proposed that considerable economic benefits could be obtained from the CEC HSR project (see Table 3). While the provincial government for potential development of HSR has preserved some parcels of land, there is still no official political commitment to undertake an HSR project in Alberta.

The 2004 study identified two possible options. One was to upgrade and share the existing Canadian Pacific Railway (CPR) line. However, this line was originally built in the late 1800s and therefore significant upgrade to the track would be needed. The second option was to construct a new line (though it would access the CPR corridor in Edmonton and Calgary) called the Greenfield which would be 184 miles (294 km) as opposed to the CPR option which would be a 194 miles (310 km).

The feasibility study of HSR of CEC in 2004 identified the need for five stations: downtown Calgary, north Calgary in vicinity of Calgary International Airport, Red Deer (a smaller city located at approximately halfway between the two much larger cities), a location near Edmonton International Airport (or south side of Edmonton) and downtown Edmonton adjacent to the Grandin LRT station and a maintenance center in proximity to Red Deer. Table 4 shows the features of the proposed service.

One unique aspect of the CEC is that it is relatively sparse in population. Currently the province of Alberta has a population just over 4 million (Government of

Table 3. Costs and net present value (NPV) of HSR in the CEC[a] (Van Horne Institute, 2004)

Representative Costs	
Total Capital Cost	$3.73 billion
Total Cost of Operations & Maintenance	$104 million per year +2%
Representative Benefits	
Net Present Value (NPV) of Government Investments	$2.7 billion for wholly public case $970 million for public-private case
NPV of Benefit to Public Safety[b]	$2.1 billion
NPV of Benefit from Reduced Emissions[c]	$107 million
NPV of Entire Project	**$6.7 million**

[a]All values in the table are based on 2009 Canadian Dollars and a 188 mph (300 km/h) technological scenario. Costs were extrapolated from the Van Horne Institute study, and were adjusted with the Bank of Canada's on-line inflation calculator to match the reference year.

[b]Includes "travel time and cost savings for system users, accident reduction and environmental benefits," Van Horne Institute, Calgary/Edmonton High-Speed Rail: An Integrated Economic Region, p. 116

[c]This value focuses solely on GHGs; it does not include CACs.

Table 4. HSR service plan – Calgary/Edmonton markets (Van Horne Institute, 2004)

	CPR	Greenfield Non-Electric	Greenfield Electric
Time/Speed	2 hr 10 min (125 mph/200 km/h) (125 mph (100 km/hr)	1 hr 46 min (150 mph/240 km/h)	1 hr 30min (200 mph/320 km/h)
Frequency (round trips)	10	10	10
Cost (round trip)-Business Non-Business	$115 $97	$115 $97	$115 $97

Alberta, 2013). Both Calgary and Edmonton have populations of less than 1.25 million each and Red Deer has a population of about 100,000. Thus compared to other regions where HSR has been implemented or where it is being proposed, the potential ridership is low. However, factors conducive to HSR include the fact that it is a wealthy corridor because of the oil and gas industry and the population growth rates are high (Government of Alberta, 2013). Projected income and employment growth rates are also high (Van Horne Institute, 2013). Also the winters are long and sometimes severe weather conditions can be dangerous for driving, thus making HSR an attractive option.

In late 2013, the Standing Committee on Alberta's Economic Future (one of three legislative policy committees established by the Legislative Assembly of Alberta) undertook a study on the feasibility of HSR in Alberta. The report was submitted in May 2014 (Standing Committee on Alberta's Economic Future, 2014). The committee heard from various stakeholders. While the creators of the TEMS and Van Horne reports made presentations, the committee also heard from individuals and groups that had a less positive view. Some of the concerns expressed included the fact that the population was insufficient and that this population was 'auto oriented' and thus there was less likelihood of potential travelers shifting modes of transport, the uncertainty of ridership forecasts and revenues, and the need for governments to subsidize not only capital costs but also operating costs (the more favorable reports had posited that ticket revenues would cover operating costs). Government subsidies in operating costs would be necessary if private investment would not be able to operate profitably without support. Further, these analyses also cast doubt on the socio-economic benefits. In fact, the Alberta Association of Municipal Districts and Counties (2010) had a report produced on the effect of the CEC HSR on rural areas and the costs to mitigate them. The committee also heard that the Alberta Government needed to develop a strategic transportation plan as well a regulatory framework, including more light rail transportation in Calgary and Edmonton, as well as a establish utility corridor between the new cities (including acquiring land) which would include space for future HSR. These were expected to support future development of HSR. The final recommendation of the committee stated that 'the Government of Alberta should not invest in a high-speed rail transit system in the Edmonton-Calgary corridor at this time because the population of the corridor is not sufficient to support the profitable operation of such a system.' The Alberta Ministry of Transportation reported that it was considering the report (Dykstra, 2014). Further, none of the candidates who recently vied for the post of Premier of Alberta (the previous Premier resigned in the spring of 2014) mentioned funding for HSR. Thus, it appears that given the report from the Legislative Assembly and the absence of support from the government leadership, it is unlikely that a HSR line between Calgary and Edmonton will be built in the near future.

INFLUENTIAL ASPECTS OF THE HSR FEASIBILITY IN CANADA

During the past two decades, Canada has experienced general erosion in passenger rail service (Hurley and Jones, 1990). Concerns regarding the current conventional passenger rail operations have concentrated on the level of capital and operating cost reductions associated with existing passenger operations. Previous attempts by VIA rail Canada and others to establish the desirability and feasibility of HSR implementation have failed to attract the interest of either the public or private sectors (Hurley & Jones, 1990).

However, the slow conventional passenger rail service has helped bring a revival of interest in HSR in recent years. The news of major investments in high-speed trains in other countries as well as the belief that HSR can help alleviate a number of socioecological problems associated with heavy fossil fuel consumption, such as climate change, air pollution, fuel dependency, congested highways, and associated social and health impacts have stimulated such interest. As mentioned, recently, EcoTrain and the Van Horne Institute have produced updated feasibility reports for HSR projects in QWC and CEC respectively. In the next sections, the influential aspects of the HSR feasibility in Canada, namely the supporting policies, HSR technology, demographic and potential ridership trends, investment costs, operation and maintenance costs, impacts on the transportation system as well as socio-economic and environmental impacts and benefits will be discussed.

Supporting Policies

In this section, we review the transportation policies in selected countries with existing HSR systems, namely France, Germany and Spain. This will illustrate how demographics, market, infrastructure, transport regulations, environmental, political and others are relevant to HSR and would be informative to the QWC and CEC proposed projects. These three countries were selected because of the availability of information. More importantly, they all have extensive experience in the development and operation of HSR networks. Though the United States (US) is seen as a country similar to Canada in many aspects, it was not considered as a comparator mainly because the development of HSR in the US is still at an early stage. The success factors and lessons learned from the supporting policies in these three countries are as follows:

First, in creating HSR service, the review of the policies shows that these three countries upgraded the system quality by building new, dedicated HSR tracks, with Germany allowing freight traffic outside of hours of operations of passenger trains.

Second, from the perspective of master planning, the three countries differed in their approach. Rather than as part of HSR networks or transportation master plans, the first high-speed lines developed in France and Spain were planned as individual projects. In contrast, the HSR program in Germany was developed as part of an overall transportation master plan including issues related to regional and town planning. Over time, there has been a growing understanding of the advantages of master planning similar to that performed in Germany. Therefore, an elaborate master plan for transport and HSR has been proven an indispensable step in the HSR network development process (EcoTrain, 2011).

Third, because of the importance of inter-modality in developing efficient transport systems, particularly with the long distance air and rail transport, the three countries all have high-speed routes with stations that are also used for other transit systems and/or are connected to international airports (in fact in some cases one can buy a combined air-and-rail ticket through the airline). Many high-speed rail routes have stations in city centers, facilitating interconnectivity with other transport modes. Thus, it appears that the inter-modality with the existing conventional railway network and interconnectivity between HSR stations and other modes of transport are two important success factors. Specifically, the factors include strong interface between HSR and conventional rail lines; locating HSR stations to provide quick access to city centers; developing intermediate stations to provide connectivity with other modes of transport; and developing light rail or commuter services linked with the HSR, including buses and airport shuttles. In the QWC HSR project, the rail connections from the HSR to airports in Montréal (Trudeau) and Toronto (Pearson) as well as the location of the proposed HSR stations at VIA rail stations downtown reflect this point.

In the CEC, while both Edmonton and Calgary airports would be connected, the VIA rail station in Edmonton is not close to the proposed HSR terminus (as mentioned, Calgary has not had conventional passenger rail service since 1985). However, this may not be an issue since the primary ridership would be between the two cities given the sparse population in western Canada. It is unlikely that many passengers from conventional rail or buses would transfer to HSR, which is the rationale for connecting other modes of transport to HSR stations. It is possible that people flying in may transfer to HSR and that would be feasible since the airports are to be connected in the plan.

The TGV in France was designed and built primarily for passenger transport. Hence, the passengers diverted from conventional rail to HSR increased the capacity for freight traffic on the conventional rail network. In contrast, the Perpignan – Figueres HSR line (also in France) has been designed and built for mixed passenger and freight traffic. In Germany, while strictly controlling the operating hours of freight trains, there is a long history in allowing freight traffic use its HSR lines. Considering the current conditions, the QWC and CEC HSRs will require building new and dedicated HSR tracks, and the increased capacity for freight transport on existing conventional lines will be a collateral benefit.

From an organizational structure perspective, in accordance with EU directives to foster in- market competition, all three countries separated the management of rail infrastructure from its operations. Following the implementation of these reforms, the HSR networks are owned and managed by state-owned limited liability companies: RFF in France, DB Netz in Germany, and ADIF in Spain. The organizations that own and manage HSR operations are SNCF in France, DB Mobility Logistics in Germany, and RENFE in Spain. Different from these European countries, the majority of the railway infrastructure in Canada has already been privatized and is currently owned by the private sector. Therefore, in order to determine the appropriate organizational structure for the HSR in Canada, before implementing similar reforms, a systematic review of transportation policies in Canada is required so as to ensure that the benefits achieved by the regulatory reforms above are maximized in Canada.

None of the three countries reviewed undertook integrated taxation policy changes directed at facilitating a transport mode shift. Operators pay usage charges in various forms to rail infrastructure providers and managers, including charges for the access right to the infrastructure and the right to reserve capacity (or the availability of routes), as well as for the actual use of train tracks, railway infrastructure and platforms. Based on the experience of these three countries, it appears that transport infrastructure requirements, environmental objectives and trends, and socio-economic considerations should be reviewed carefully to evaluate whether there would be a need for a comprehensive taxation policy for achieving the desired transport mode shift and desired levels of self-financing for the upcoming HSR projects.

However, the environmental problems were not key issues considered in the initial stages of the HSR system development in these European countries, environmental concerns are an important factor today. Over the years, additional steps have been taken to reduce the negative environmental influences related to the construction and operation of HSR systems by undertaking environmental assessments of new rail lines, reduction of noise levels, and using environmentally friendly energy. In addition, for traction technology selection, the trend is to reduce dependence on fossil fuels as the main source of energy. The availability of nuclear electricity allows for electrical traction for HSR. In Canada, while most of the electricity in Quebec and Ontario is generated through hydroelectricity, in Alberta over 40% is generated from coal (Government of Alberta, 2014). However, it is expected that by 2032, only 12% of Alberta's power will come from coal. The rest would be primarily from natural gas and, to a lesser extent, from wind. Thus, the energy sources for electric HSR would be more environmentally friendly.

Throughout Europe, the HSR networks are electrified, using standard gauge, and will eventually all use the European Rail Traffic Management System (ERTMS) for signaling, because one of the major policy objectives in the EU is interoperability. However, this may not be necessary for the HSR lines planned across North America now, careful consideration should be paid to the development of interconnected, compatible networks within Canada and with the U.S in the long term.

Based on the funding policy for HSR in these three European countries, the HSR infrastructure funding primarily comes from governments and the HSR operators collect the usage charges. HSR fares are not subsidized, except for regional services targeting passengers or specific users who can benefit from subsidized fares (i.e. veterans, pensioners and students). The pricing of long distance HSR services are usually based on the market and primarily determined by competition from airlines and commercial goals, with a view to making HSR operations profitable. Considering the context in Canada, further investigation is required to assess optimal approaches that could be applied to the funding of HSR infrastructure and operations.

HSR Technology

The HSR technology includes all the civil, mechanical, electrical, electronic, and informational components of an HSR service, together with all equipment and processes needed to operate a HSR service safely and cost effectively on a regular basis (EcoTrain, 2011). It also refers to consideration of the interactions between each of these components as well as with their environment (EcoTrain, 2011).

Clearly, technology is definitely one of the principal influential aspects of the HSR system since the HSR services require a high level of compatibility between their infrastructure, rolling stock, and other equipment which determines the performance levels, quality of service, safety and cost.

In this section, the technical base for the HSR system in Canada will be discussed from seven aspects, namely rolling stock, track, energy, signals and telecommunications, track sharing and system protection as well as winter operation.

Rolling Stock

Typical high-speed trains are generally bi-directional sets with a control cab at each end of the train for quick direction changes at the end of the line or terminal. The trains have semi-permanently coupled cars with fixed configurations during operation. They are also equipped with couplers at both ends, which can be used for coupling a second trainset or for coupling with conventional rolling stock in an emergency, such as towing a stranded train.

There is a wide variety of rolling stock options for HSR in Canada with maximum operating speeds ranging from 125 to 220 miles/hr (200 km/h to 350 km/h). The choice of the rolling stock among different operating speeds is directly based on the design speed of the line on which they run and the distances between the cities, they serve. Therefore, according to the feasibility report by EcoTrain, in terms of travel quality and performance, investment and recurrent expenses, as well as environmental impacts, the proposed two representative technologies in the QWC are mainly based on minimum operational speeds of 125 miles/hr (200 km/h) and 185 miles/hr (300 km/h). The choices include diesel traction for 200 km/h operation and electric traction for 300 km/h operation. For the QWC, the two options are designated as F200+ and E300+ respectively. Since past studies have proved that it would be prohibitively expensive to provide adequate ventilation in the Mount Royal tunnel in Montreal to accommodate diesel-powered locomotives, it is noteworthy that the F200+ option would require dual-mode locomotives with electric and diesel tractions on the Quebec City-Montréal segment.

Similarly, for the CEC HSR project, the 2004 study by Van Horne Institute has assumed that travel time of two hours or less was the threshold required to attract sufficient demand supporting the HSR service. To meet this threshold, technology options operating at a speed of 200 km/h or more are necessary. Hence, the two representative technologies suggested in the 2004 study are the 125-150 mph (200/240 km/h) JetTrain (an experimental high-speed train for passenger transport created by Bombardier Transportation) and 200 mph (320 km/h) electric trains. Considering the mixed freight traffic operations, the CPR option design was on the basis of 200 km/h operation while the Greenfield option was designed for both a 240 km/h JetTrain and a 320 km/h electric train.

Currently, the length of high-speed trains in operation varies between 55 metres (m, 182’) and 405m (1337’) which mainly depend on the seating capacity needed, the length of station platforms, etc. The standard length is typically 200m (660’) or 400m (1320'). Typically the 200m trainsets with six to eight coupled cars or eight to eleven articulated cars are chosen for moderate passenger traffic volumes and are widely used in Europe, China, and Japan. The capacity of this 200m long high-speed trainset with single-deck cars is about 400-450 seated passengers. Two of these trainsets could be coupled in operation to double the train's length and seating capacity so as to meet seasonal or daily peak demand, with the appropriate platform length provided.

To improve the service quality, most HSR cars are designed for level boarding and alighting at stations, thus facilitating the accessibility for disabled people, and reducing the time for passenger entry and exit and enhancing passenger comfort.

In the feasibility report by EcoTrain for the QWC HSR system, for both F200+ and E300+ technologies, the proposed the high-speed train would be 200m long comprising six to eight coupled cars, or eight to ten articulated cars, with a capacity of 400 passengers per trainset. A train would consist of one or two trainsets, depending on the capacity required. As this is a kind of HSR train widely used in the countries with existing HSR systems and also suitable for QWC HSR project, it can also be a potential option for the CEC HSR system.

Track

The track provides the trainsets with physical guideway to run on under the desired conditions of speed, comfort and safety. In principle, HSR could use any track. However, the tracks must have the proper gauge, wheel/rail compatibility, compatible signaling system and energy supply suitable for operation of the trainsets. Except for the track technology, alignment parameters namely curve radii, banking and allowable banking deficiency for the trainsets, as well as the quality of track geometry, are the limitations to the maximum allowed speed.

The two common options for track configuration are track with ballast and ballast-less track systems. Ballasted track is the conventional method for installing railway tracks. The main advantages include highly mechanized construction procedures, low sensitivity to construction defects, good adaptability to settling and low noise emission. However, for the ballasted track, the operational loads may lead to changing of the initial geometric conditions. Ballasted track construction also has higher monetary and time costs for inspection and maintenance as well as heavier dead loads on bridges and other structures. Ballast-less tracks, on the other hand, replace the ballast with concrete in the form of prefabricated or cast-in-place concrete slabs aimed at preserving the initial geometric conditions of the track. Therefore, ballast-less track construction has lower monetary and time costs for inspection and maintenance, but results in higher construction costs due to the required accuracy. Ballast-less tracks also require good sub-soil bearing capacity. Today, various methods of ballast-less track construction are available.

Based on the two feasibility studies, both ballasted tracks and ballast-less tracks are suitable for HSR operations: the choice should be made at the design stage, according to the bearing capacity of the sub-soil. Specifically, in the report by EcoTrain for the HSR in QWC, considering the lack of detailed information on sub-soil conditions, a 50%-50% combination of ballasted and ballast-less track was chosen as the representative technology for construction and maintenance costs forecasting. Both systems are proven for operation at speeds higher than 300 km/h. There are no relevant specification changes within the 200 up to 350 km/h speed range.

Energy

There are mainly two kinds of energy sources for HSR - diesel fuel and electricity. The choice between these two systems should be based on the requirements of the intended operation. Little infrastructure is required by diesel power. However, there is a limitation that the diesel powered traction can only be installed in a typical 200m long diesel-powered trainset, with a maximum traction power of 4,000 kW. Hence, for HSR service, diesel traction can be used only in the lower speed range.

Electricity makes it possible to power more powerful trainsets. For a typical 200m long trainset, the current maximum traction power installed is near 10,000 kW. Electric power is suitable for the entire speed ranging from 200 km/h to the maximum of 350 km/h. However, it will lead to higher infrastructure costs when compared with the diesel traction option.

The respective advantages and disadvantages of diesel and electric traction are summarized in Table 5.

Table 5. Comparison of diesel and electric energy sources for HSR (EcoTrain, 2011)

Characteristic	Diesel Power	Electric Power
Financial		
Investment costs	Lower due to absence of power supply and distribution systems	Higher due to power supply and distribution systems
Operating costs	Higher traction energy costs	Lower traction energy costs
Maintenance costs	Higher for vehicle due to complexity of power apparatus	Higher for infrastructure due to poor distribution system
Operational		
Maximum speed	200-225 km/h	>300 km/h km/h

continued on following page

Table 5. Continued

Characteristic	Diesel Power	Electric Power
Maximum installed power	4500 kW	10,000 kW
Flexibility and autonomy	Can be rerouted to non-electrified lines in case of power loss	Cannot be rerouted to non-electrified lines in case of power loss
Clearance	No restrictions	Restrictions due to catenary wires and Poles
Circulation in tunnels	Requires ventilation and may be disallowed	Does not require ventilation
Energy consumption	Braking energy may not be recovered	Braking energy may be recovered
Weather susceptibility	Emergency on-board power supply from main diesel engine	Emergency on-board power supply for winter operation is necessary
Environmental		
Greenhouse gases (GHG)	Direct emission of GHG's	Indirect emission GHG's where electricity is produced from thermal plants
Criteria air contaminants (CAC)	Direct emission of CAC's	Indirect emission of CAC's where electricity is produced from thermal plants
Danger	Large amount of flammable liquids on board	Minimal amount of flammable liquids on board
Noise	Higher, especially at low speeds	Lower at low speeds
Source of energy	Constant dependence on fossil fuels	Dependence on fossil fuels only if from thermal power plant

Hence, diesel power is usually applied to lines where the ridership and relative traction power requirements are low while electric traction is applied to busy lines with high passenger volumes and relative high traction power requirements (e.g., higher speed). Since HSR trainsets often have very high traction power requirements, electricity is usually the preferred power source.

The evaluation of the two main energy sources, diesel fuel and electricity should be based on their construction, operation, and maintenance costs, their environmental impacts, as well with respect to the financial and economic analysis. In the CEC HSR projects, for the 200 km/h option, diesel power can be chosen while for the higher speed options, the preferred energy source should be electricity. Since the distance between Calgary and Edmonton is only about 300 km surveys indicated that while a travel time of two hours was seen as desirable, there was not much utility in reducing it to ninety minutes. Thus a 240 km/h train would be sufficient.

Signals and Telecommunications

Collision protection, speed control, protection of interlockings, and protecting train routes on the line and in stations are the basic functions of a signaling and train control system. For conventional trains (up to 160 km/h), the railway signal systems first detect whether a track section is occupied and then relay that message to wayside signal lights displaying the status of track sections ahead.

For higher speed trains (200 km/h and more), an automatic, cab-based signaling system needs to be installed as the observation of the wayside conventional signals at higher operating speeds, and in inclement weather (e.g., rain, fog or snow) is not reliable. The current railway signaling system used for conventional railway in Canada cannot meet the needs of HSR operating at the speed above 200 km/h, nor can it be upgraded for such an operation. Therefore, for both QWC and CEC HSR systems, instead of wayside installation for conventional signals, a modern railway signaling system with signals information displaying inside the cab will need to be installed. Further, HSR trainsets would also need to have the necessary equipment to interface with the signaling systems currently in operation in the case of shared tracks with commuter trains. Similarly, the commuter rail trains will need to be equipped with on-board devices to interface with the HSR signaling system.

A severe hazard to both rail and road traffic are grade crossings. Currently, for the safety of the HSR system, grade crossings with roads have been eliminated from almost all the railways in European countries wherever the train speed exceeds 160 km/h. In Asia, since most new railways are designed for HSR, the grade crossings are avoided from the beginning. Therefore, for QWC HSR project, it is assumed that there will be no grade crossings of HSR tracks with other tracks, main and secondary roads, or private farm crossings. Similarly, for CEC HSR system, it is assumed that 72 public level crossings will be closed and 46 others grade separated.

For safety and security, most high-speed railways are also segregated from adjacent areas (i.e., access controlled). In the feasibility report for both the QWC and CEC HSR systems, it was assumed that the whole route would be fenced.

Winter Operation

Operating HSR during winter has been recognized as a critical issue ever since the first studies of HSR in Canada were conducted. It is anticipated that the extreme climate may lead to various design and operational issues.

First, extreme weather conditions may cause problems on the track and infrastructure such as frost in sub-soil, failure of turnouts, snow accumulation and ice or ballast thrown on platforms. Second, ice formations falling from the train or air turbulence caused by a fast moving trainset can cause uncontrolled movement of ballast stones. This is called "ballast pickup" or "flying ballast". Furthermore, the failure of high voltage electric transmission lines, the failure of one or several power sub-stations supporting the HSR line, and/or some part of the overhead contact wire line being covered with ice are major problems that can influence the performance of the electrified power system during winter in the two corridors. While the QWC has weather events such as ice storms, the CEC can experience temperatures below -40C.

With the advancement of the technology over the years, solutions to the infrastructure problems, ballast pickup and the power system problems listed above have been developed. Canada has rich experience and technical expertise on transportation and distribution of electricity under severe climate conditions. This expertise will be helpful in the design of the HSR system for QWC and CEC.

Signaling systems may also have problems in the winter operation. When a radio-based cab signaling system is applied, there is no need for a new wayside signaling system. In the HSR system, only robust installations such as rail bonds, connectors, transponders and possibly axle counters should be used as they are sealed and duly protected against snow, ice and very low temperatures and can work reliably under severe winter conditions.

Besides Japan and South Korea, HSR winter operation has also been proven in Finland and Sweden where winter conditions are similar to that in Eastern Canada with respect to snow accumulation and temperature. Both Finland and Sweden have been operating 200+ km/h electric trains for many years. South Korea, with an average temperature of -6°C and a low temperature of -15°C in winter, operates HSR at 300 km/h. In Russia, another location with severe winter conditions, the Moscow - St. Petersburg HSR line is being upgraded to a maximum operating speed of 250 km/h and eventually to 300 km/h.

Since the accumulated research, development, and experience in HSR winter operation are considerable now, the winter operation of HSR systems in both QWC and CEC should not be a major problem. In the initial design, special winter weather conditions such as low temperatures, snow and ice accumulation should be evaluated with a view to minimize their impact on HSR operation.

Demographic and Potential Ridership Trends

The demographics in different corridors and the potential ridership trends are two important aspects which will influence HSR development, especially from a revenue perspective.

Demographic and Potential Ridership Trend in QWC

In the demographic analysis of the feasibility report by EcoTrain, it was noted that in the 1992 Québec Ontario High Speed Rail Project Study (QOHSRPS) the predicted population in 2006 was overestimated. Similarly, the report noted that the forecasted population for 2025 was too optimistic (see Table 6). The QOHRPS study also overestimated the annual growth rate of trips by auto and air, and underestimated trips by rail and bus (see Table 7). According to a study by Hurley and Jones (1990), it was assumed that the growth in HSR traffic would come at the expense of car travel while the percentage of bus and air travel would not be significantly impacted.

In addition to the changes in the population and the annual growth rate of trips by different transport modes, land use modification, travel patterns and environmental requirements for the different modes of transport will also influence the forecast for HSR travel demand and hence the revenue. Specifically, between 1992 (previous study) and 2011, real estate development had necessitated modifications to HSR route alignments. Large increases in commuter rail services in Montreal and Toronto have induced changes in travel patterns. Also during the past twenty years, there also have been updates to environmental requirements.

The travel demand forecasting in the feasibility report by EcoTrain for the QWC HSR project (using a nested logit-based mode choice model) was based on the following independent variables and assumptions: QOHSRPS results, travel data by

Table 6. Comparison of population forecasts – EcoTrain Study versus the QOHSRPS study (EcoTrain, 2011)

Year	QOHSRPS Study Population Forecast (thousands)				
	1992	**2005**	**2006***	**2025**	
Quebec	6826	7800	7845	8700	
Ontario	10262	12200	12375	15700	
Total	17188	20000	20220	24400	
Year	**EcoTrain Study Population Forecast (thousands)**				
			2006	**2025***	**2031**
Quebec			7546	8501	8802
Ontario			12161	15024	15929
Total			19707	23525	24731
Difference			-2.5%	-3.6%	

*Interpolated

Table 7. Comparison of total trips growth 2006-2025 — EcoTrain study versus the QOHSRPS study (EcoTrain, 2011)

	Trips (millions) 1992	1992-2005 Total Growth	Trips (millions) 2005	2005-2025 Annual Growth	Trips (millions) 2006	Trips (millions) 2025	2006-2025 Total Growth	Trips (millions) 2031
QOHSRPS Auto Air Rail Bus	 99.0 4.1 2.9 2.6	 33% 46% 0% 0%	 131.2 6.0 2.6 2.6	 2.10% 2.58% 0% 0%	 134.0 6.1 2.9 2.6	 198.8 10.0 2.9 2.6	 48% 62% 0% 0%	
Total Trips	108.6	31%	142.7	2.05%	145.6	214.3	47%	
EcoTrain Auto Air Rail Bus								
Total Trips					92.7	106.8	15%	117

Notes:

- The 1992 Study included additional markets such as Hamilton to Toronto so total trips are not comparable, but growth are comparable since the 1992 Study growth applied to all markets.
- To permit easier comparison, trips were interpolated for 2006 in the 1992 study and for 2025 for this study.

air, rail and bus in 2006 provided by Transport Canada, travel surveys in 2009 conducted by EcoTrain and HSR fare structure designed to maximize revenue rather than ridership. The travel demand forecasting for QWC HSR by EcoTrain indicates that:

- The ridership of HSR systems mainly comes from passengers shifted from other transport modes and the travel demand induced by HSR services. On the basis of the traveler preference survey conducted by EcoTrain, among the current VIA Rail passengers, 50% of them would like to use HSR under certain conditions. Current users of auto, air and bus modes would generally continue to use these modes, however, 19% of auto passengers, 40% of air passengers and 32% of bus passengers are willing to select HSR as their second preferred choice of transport under certain conditions. Besides, it is estimated that the QWC HSR system will induce 800,000 passengers (new travelers generated by the availability of HSR, who did not exist before).
- After the introduction of QWC HSR system, auto will still be the dominant travel mode within the QWC holding about 85% share of the market, but HSR will become the transport mode taking up the second largest market share of close to 9%. As VIA will be replaced by the faster but more expensive HSR, the cost sensitive VIA passengers will change to bus travel, the cheaper mode.

- In total, the E300+ option will attract 11.1 million passengers and generate a revenue of $1.3 billion (2009 CAD) while the F200+ option will attract 10.2 million passengers and generate a revenue of $1.2 billion (2009 CAD) in 2031 for the entire corridor (EcoTrain, 2011). Particularly, the E300+ option will attract more business travelers as compared with the F200+ option.
- The origin and destination of about 80% of the trips made in HSR are expected to be in one of the seven urban areas within the QWC, hence making up the main market for HSR. Specifically, the Quebec-Montreal-Ottawa-Toronto segment is expected to generate nearly 80% of the total HSR travel demand in QWC, and the Montreal- Ottawa-Toronto segment will generate a proportion of just over 55%.

Table 8 shows the expected annual ridership in thousands and the annual revenues in thousands of 2009 Canadian dollars for the whole QWC, as well as for other segments in the year 2031 and 2041 for the F200+ option and E300+ option respectively. It indicates there is growth in every segment of the corridor.

Table 8. Expected yearly ridership and revenues (in thousand 2009 CAD) (EcoTrain, 2011)

Trip Purpose and Segments	Technology	2031		2041	
		Trips (x1000)	Revenue (x1000)	Trips (x1000)	Revenue (x1000)
Business	F200+	5,140	$699,000	5,634	$774,000
	E300+	5,659	$791,000	6,202	$875,000
Non-Business	F200+	5,088	$509,000	5,496	$547,000
	E300+	5,465	$562,000	5,902	$605,000
Total Quebec City – Windsor	**F200+**	**10,227**	**$1,208,000**	**11,130**	**$1,353,000**
	E300+	**11,124**	**$1,321,000**	**12,105**	**$1,480,000**
Quebec City – Toronto	F200+	7,971	$985,000	8,609	$1,072,000
	E300+	8,885	$1,119,000	9,604	$1,219,000
Montréal – Toronto	F200+	5,681	$722,000	6,234	$798,000
	E300+	6,346	$825,000	6,969	$913,000
Toronto – Windsor	F200+	2,169	$189,000	2,410	$211,000
	E300+	2,135	$193,000	2,370	$215,000

Demographic and Potential Ridership Trend in CEC

In 2008, TEMS/Oliver Wyman completed an in-depth analysis of the potential market suggested by the previous HSR study of CEC in 2004 which only included a preliminary assessment of potential ridership and revenue. The preference and usage survey data collected in this analysis were used as the input in the predictive models in the latest feasibility study of CEC HSR system by Van Horne Institute in 2013. This data was based on population, employment, and income and economic projections, including changes in traffic congestion and fuel prices. The main results of ridership and revenue forecasts in this study are as follows (see Table 9):

According to the sensitivity analysis carried out on the model results, demographic change is one of the three most critical variables that influence the CEC HSR ridership and revenue. The other two critical factors are congestion levels and gasoline prices. For the demographic factors, there are three scenarios included in the TEMS model for population, employment and average household income growth at five-year intervals from 2001 to 2051.

Specifically, comparing the current population projections by the Government of Alberta - Treasury and Finance department to those used in the TEMS ridership forecasting, it shows that the growth of population is expected to be more substantial than the previous estimate (see Figure 3). For the employment forecasting in the CEC, Figure 4 reveals that the employment in the current projections for 2011 met the TEMS high forecast for this year, and the employment growth trend is significantly stronger than that assumed in the TEMS ridership forecasting (The current

Table 9. Expected CEC HSR ridership and revenues (Van Horne Institute, 2013)

Corridor Ridership (Millions)								
Ridership	**200 km/h**		**240 km/h**		**320 km/h**		**480 km/h**	
Year	**2021**	**2051**	**2021**	**2051**	**2021**	**2051**	**2021**	**2051**
Worst case	1254	1860	2034	2839	3359	5068	4766	7180
Base case	1554	2821	2518	4301	4136	7657	5816	10745
Best case	2207	4618	3583	7058	5615	11947	7897	16751
Corridor Revenues (in Million 2006 CAD)								
Revenue	**200 km/h**		**240 km/h**		**320 km/h**		**480 km/h**	
Year	**2021**	**2051**	**2021**	**2051**	**2021**	**2051**	**2021**	**2051**
Worst case	60.6	90.2	105.3	177.1	204.7	353.3	489.3	902.8
Base case	75.2	137.1	156.7	269.0	328.2	609.9	608.0	1127.9
Best case	119.2	250.6	223.0	441.7	485.3	1035.7	825.6	1758.7

Figure 3. Corridor population projection comparison (Van Horne Institute, 2013)

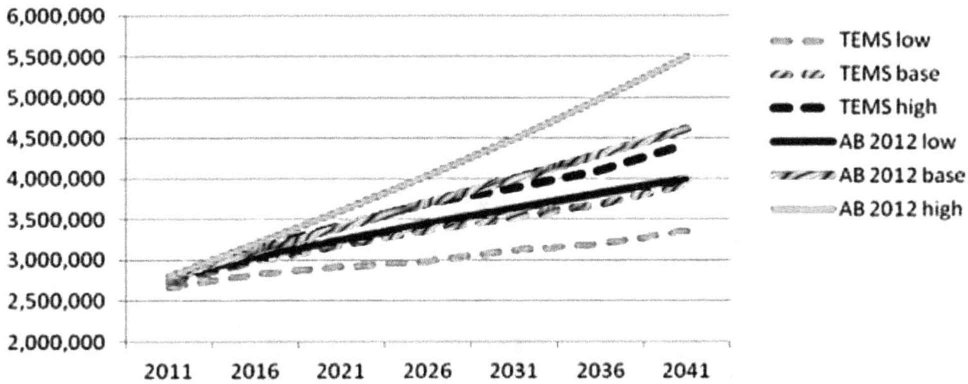

Figure 4. Corridor population projection comparison (Van Horne Institute, 2013)

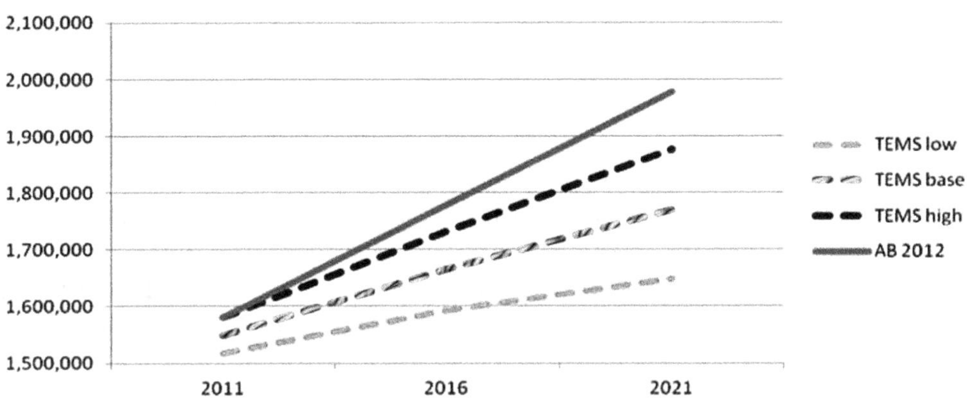

projection only provides part of the TEMS estimation range until the year 2021.) The percentage of growth in average household income forecasted by TEMS was compared to the data reported by Statistics Canada from 2001 to 2010, because Alberta Treasury and Finance does not forecast average household incomes. The results are illustrated in Figure 5, which shows that, compared with the estimation by TEMS, significantly stronger income growth is expected to occur in Red Deer, while in both Calgary and Edmonton, the income growth is fairly comparable to the high estimation by TEMS.

On the basis of the updated socio-economic and other factors (e.g., fuel cost and traffic congestion) impacting ridership in the TEMS model, the TEMS high demographic forecast without adjustment for fuel or traffic congestion is the best fit. The latest analysis by Van Horne Institute in 2013 suggests that a higher estimate in accordance with the TEMS high demographic forecast is more appropriate for

Figure 5. Average household income % growth 2001 to 2010/11 (Van Horne Institute, 2013)

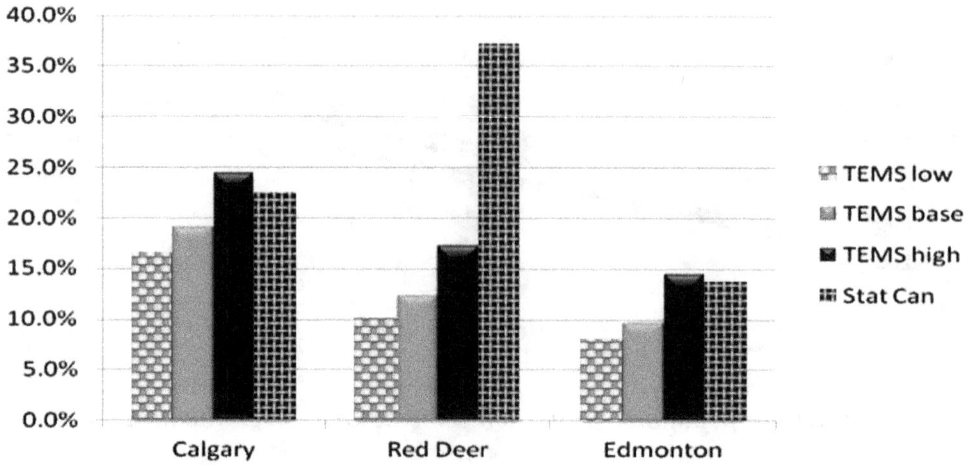

ridership forecasting, which is 24% to 34% higher than previously estimated (see Figure 6).

Due to the absence of updated stated preference surveys, the updated revenue forecasts in the latest report by Van Horne Institute were based on the updated average fare derived from the TEMS base forecast updated to 2013 dollars using the Consumer Price Index (CPI), which result in expected revenues in 2031 ranging from $138 to $1,124 million (2013 CAD) and from $217 to $1,785 million (2013

Figure 6. Comparison of ridership forecasts (millions) (Van Horne Institute, 2013)

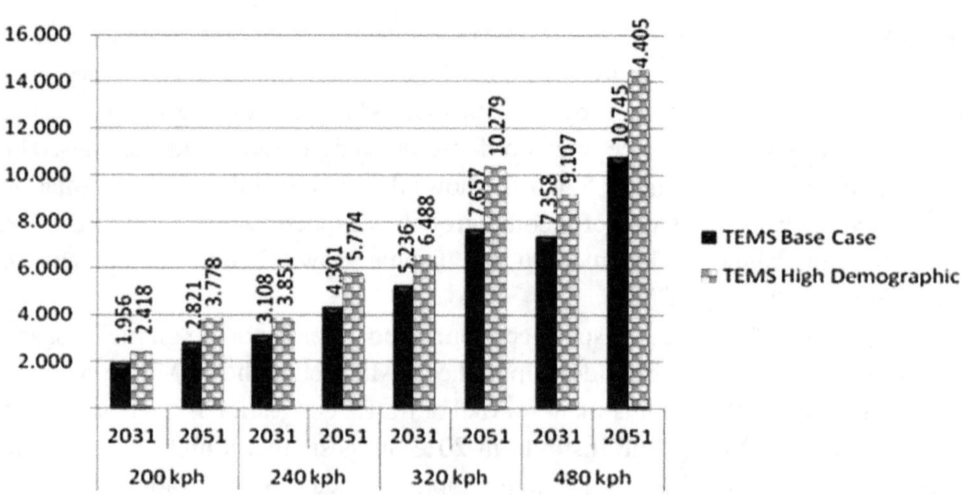

CAD) in 2051 according to the technology used (e.g., different HSR service speed). It is noteworthy that the revenues in 2051 are 58% higher than previously predicted. The results are illustrated in Figure 7.

Costs of HSR System

The costs of the HSR system undoubtedly have a great impact on its feasibility. The costs can be divided into two parts: the investment costs and operation and maintenance costs.

Investment Costs

On the basis of the technologies, routes, stations and services designed for the HSR systems, investment costs are estimated through the following aspects, namely conceptual design, preliminary and detailed engineering, construction supervision and project management costs, construction costs as well as acquisition costs of property, rolling stock and rail system.

Hurley and Jones (1990) point out that HSR is much more capital intensive than either conventional rail or other transportation forms. Capital costs consist of two types – equipment and infrastructure, with the latter being more expensive. Further,

Figure 7. Projected revenue (in constant million 2013 CAD) (Van Horne Institute, 2013)

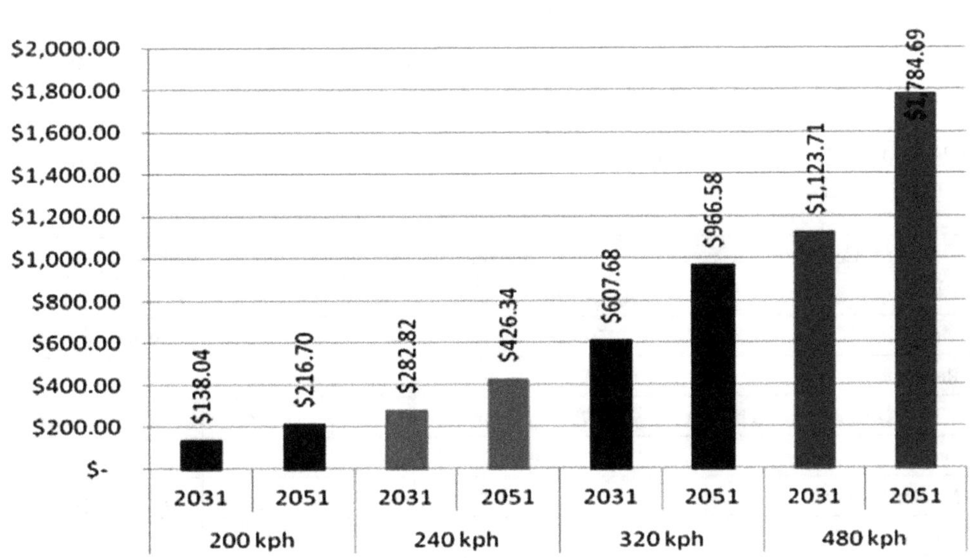

they point out that infrastructure costs are lumpier. For example laying a double track is much more expensive than laying a single track.

For the QWC HSR system, the total investment costs for the whole corridor and other segments estimated by EcoTrain and the previous QOHSRPS are illustrated in Table 10.

The differences in the expected investment costs above can be explained from the different assumptions. First, the difference between the two technologies is primarily caused by the costs of the different power supply equipment (diesel or electricity). Grade crossing is another aspect that results in a cost difference between the two technologies. In the report by EcoTrain, instead of keeping many grade crossings open in the F200+ option, the elimination of all grade crossings reduced the cost difference between the two options.

As an important element of the total investment cost, differences in the HSR construction costs have a powerful impact on the total investment costs in EcoTrain and QOHSRPS. Specifically, in the report by EcoTrain for the QWC HSR project, it is assumed that the HSR will run on double tracks with new track material along the whole corridor. EcoTrain also included a 70% probability of meeting expected project management and contingencies costs, while the probability is only 45% in the QOHSRPS. Further, separating freight transport from HSR by using different tracks or setting different operating hours can also influence the construction costs. There are other factors that can result in the investment costs increase in the EcoTrain study compared to the QOHSRPS, such as avoiding piers in rivers, using the ballast-less track instead of ballasted track to minimize maintenance costs where soil conditions permit, upgrading the signaling system and so on.

To help evaluate the investment costs of the HSR systems in Canada, the EcoTrain's estimates for the QWC E300+ HSR project are also compared with actual HSR costs in other countries. According to an extensive study by the BBVA Founda-

Table 10. Expected total investment costs for the QWC HSR project (in million 2009 CAD) (EcoTrain, 2011)

Cost Item	F200+			E300+		
	QOHSRPS	**EcoTrain**	**Ratio**	**QOHSRPS**	**EcoTrain**	**Ratio**
Infrastructure	7501	9723	1.30	8972	10268	1.14
Rail systems	8477	9205	1.09	8924	11032	1.24
Total Quebec City -Windsor	**15978**	**18927**	**1.18**	**17896**	**21300**	**1.19**
Quebec City -Toronto		13983			15987	
Montreal -Toronto		9067			10952	
Toronto -Windsor		4919			5614	

tion (de Rus, 2009), the lowest project costs were on the Paris-Lyon TGV (France) through generally flat terrain and the Madrid-Lleira AVE (Spain) through varied terrain at $8.1 million CAD (2009) per kilometre and $13.4 million CAD (2009) per kilometre respectively, while the most expensive was the Bologna-Florence AV (Italy) at $113.4 million CAD (2009) per kilometre. Excluding property acquisition, planning, and rolling stock costs, the E300+ representative HSR project investment cost is expected to be $12.8 million CAD (2009) per kilometre, which is about 60% more than the Paris-Lyon TGV and about the same unit cost as the Madrid-Lleira AVE (EcoTrain, 2011).

The reason that the QWC HSR estimated investment costs are on the low side of the actual costs of 24 different HSR projects can be explained by the relatively favorable terrain and topography in the corridor and the availability of railway Right of Way (RoW) in the major cities.

In the up to date CEC HSR study by Van Horne Institute in 2013, the investment costs are generally analyzed from three parts: property, rolling stock and other capital costs. The property costs include land acquisition costs for proposed stations and right-of-way. Assuming minimal land acquisition in downtown Calgary and Edmonton as well as the approaches to both cities, the current cost estimated for the five proposed stations and maintenance facility (near Red Deer) land acquisition was expected to be approximately $31.7 million (2013 CAD) based on a survey of current commercial land sale price listings and recent transactions.

The analysis of the land acquisition for RoW in the 2004 study suggested the requirement of 108 to 309 hectares (including land severances) for the CPR option, 1,490 hectares for the Greenfield options as well as 1,211 hectares in land severances (Van Horne Institute, 2004). Using Statistics Canada's data on the increase of farm land values from 2004 to 2010 (67.76%) and Farm Credit Canada's data from 2010 to 2013 (21.4%) as the inflation rates, the updated land acquisition costs estimation for RoW in CEC HSR project was obtained. The current estimates of total property acquisition cost for different options are shown in Table 11.

The rolling stock procurement cost is another important part in the analysis of the CEC HSR project investment cost forecasting. The estimation of the procurement cost was based on two approaches. First, since the inherent limitations of the

Table 11. Property cost estimate (stations, maintenance facility & RoW) (Van Horne Institute, 2013)

Options	$ 2004 (millions) CAD	$ 2013 (millions) CAD
CPR option	$22.8	$35.3
Greenfield option	$47.8	$82.3

CEC HSR rolling stock procurement were analyzed thoroughly in the previous study, in the latest report by Van Horne Institute, one approach was to update the 2004 rolling stock procurement cost estimate by using the Canadian inflation rate of 18% (since 2004). Second, based on the order size of the CEC HSR study in 2004, rolling stock procurement contracts for the speed of 200 to 250 km/h and the speed of 320+ km/h awarded to the three largest train manufacturers (Alstom, Bombardier and Siemens) between 2004 and 2013 were reviewed. The results were inflated to 2013 CAD using an adjustment factor for North Americanization. The Table 12 below shows the comparison these two approaches. The cost for rolling stock procurement in the latest study was assumed to range from $ 390 to $445 million (2013 CAD) for the options.

Besides the costs of the rolling stock, estimates for other capital costs of the CEC HSR system in the latest study were based on the initial work prepared in 2004 and inflated over the construction period. The estimated capital costs for the three different options are shown in Table 13. It is suggested that the CEC capital cost forecasts for each option are: $ 2.6 billion for the CPR option, $ 3.9 billion for the Greenfield non-electric option and $ 5.2 billion for the Greenfield electric option.

Similar to the analysis in the QWC HSR study by EcoTrain, a comparison between the CEC HSR investment costs estimate and actual HSR costs in other countries is also recommended as it will also be useful for the evaluation of the CEC HSR system.

Operations and Maintenance Costs

Apart from the investment costs, the estimation of operation and maintenance (O&M) costs is also crucial in the costs analysis for the feasibility of the HSR project. In the analysis of the QWC HSR system O&M costs by EcoTrain, cost elements related to operations are: HSR train operation, including the cost of salaries and supplies for train crews, energy consumption and the operational control centre (OCC);

Table 12. Rolling stock cost estimate for the CEC HSR system (in millions 2013 CAD) (Van Horne Institute, 2013)

5 Train Sets	Inflated 2004 estimate	Supplier Based Estimate
Up to 250 km/h non-electric (8 cars)	387	N/A
(10 cars)	438	N/A
Up to 250 km/h electric (8 cars)		338
(10 cars)		348
320 to 380 km/h electric (10 cars)	444	443

Table 13. Comparison of CEC HSR capital costs (in million 2013 CAD) (Van Horne Institute, 2013)

	CPR (200 Km/h)	Greenfield (250 Km/h)	Greenfield Electric (320 Km/h)
Property	35.3	82.3	82.3
Rail infrastructure	1073.9	2048.9	2048.9
Stations and Parking	39.5	39.5	39.5
Maintenance facilities & equipment	96.6	107.4	112.0
Road work	398.5	175.3	175.3
Electrification	0	0	888.7
Total construction costs	**1643.9**	**2453.5**	**3346.8**
Vehicles, tvms & Communication Systems*	**409.4**	**481.9**	**491.8**
Engineering	126.9	227.3	313.3
Project management	13.9	137.4	190.1
Testing and commissioning, ops prep	11.7	11.7	11.7
Insurance & bonding	249	88.3	116.5
Total engineering & management	**177.4**	**464.7**	**631.6**
Contingencies	346.0	525.3	716.1
Total project costs	**2576.6**	**3925.4**	**5186.3**

*TVM-Ticket Vending Machine

customer services, including the cost of salaries and supplies for on-board service staff, station operation, ticketing and booking sales, as well as advertising; and other recurrent costs, including insurance and claims, property and sales taxes, and right of way and usage fees paid to railways.

The cost elements related to maintenance included in the analysis by EcoTrain are rolling stock and infrastructure. Rolling stock maintenance costs include the cost of salaries and supplies for personnel performing routine maintenance, cleaning and maintenance, inspection, and minor and major revisions. Infrastructure maintenance costs include the cost of salaries, supplies for the staff, the cost of maintenance equipment operation, maintenance for stations and shops, track, energy supply and distribution, bridge and other structures, as well as signaling and telecommunication system. The estimated total O&M costs of the entire QWC HSR system for the first full year of operation (2025) are shown in the following Table 14:

For the analysis of the O&M costs in the CEC HSR system, fuel cost was considered as an important component. In the latest feasibility report by Van Horne Institute, the recent fuel cost was analyzed in both crude cost per barrel and gas pump fuel price per liter, through comparing the estimates of three scenarios based on West Texas

Table 14. Total operations and maintenance costs (in million 2009 CAD per year) (EcoTrain 2011)

Cost Item	F200+	E300+	Ratio E300+/F200+
Train operations	84.7	86	1.02
Customer services	123.9	134.2	1.08
Equipment maintenance	110.6	87.0	0.79
• Infrastructure • Maintenance	54.5	92.4	1.70
• Management and other • Costs	118.1	120.5	1.02
Total per year	**491.8**	**520.2**	**1.06**

Intermediate Crude (WTI) by TEMS and the figures from Government of Alberta - Treasury and Finance in the 2012 Fiscal Plan (AB 2011 and AB 2012 in Figure 8) and Natural Resources Canada (NRC 2011, NRC 2012 in Figure 9).

The comparison results show that the fuel cost per barrel is close to the TEMS high scenario. The average gas pump fuel price per liter is less than the high scenario level of $1.50 per liter, but slightly higher than the base scenario.

In the latest Van Horne Institute study for the CEC HSR system, the operating costs were estimated by updating the base VIA Rail model in the Van Horne Institute, (2004) study with appropriate increases. More specifically, the labor (VIA Rail) and diesel fuel costs have increased more than inflation (CPI) by approximately 26% to 29% and 87.9% respectively while the electricity costs have increased by only 4%. The costs of on-board products such as food are inflated by CPI and adjusted

Figure 8. West Texas Intermediate crude cost per barrel in USD (Van Horne Institute, 2013)

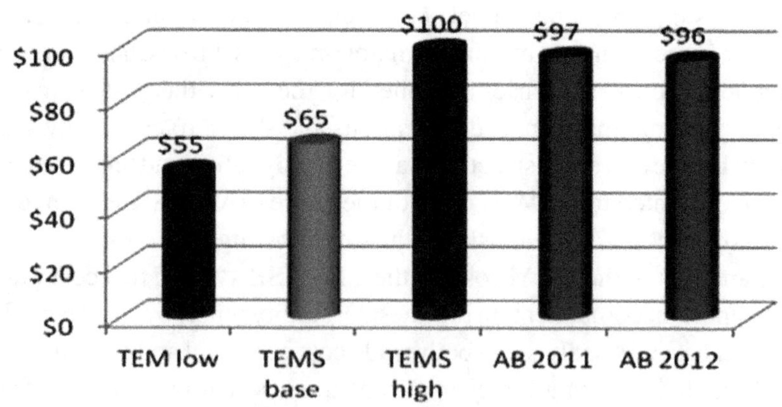

Figure 9. Gas pump fuel price per liter (Van Horne Institute, 2013)

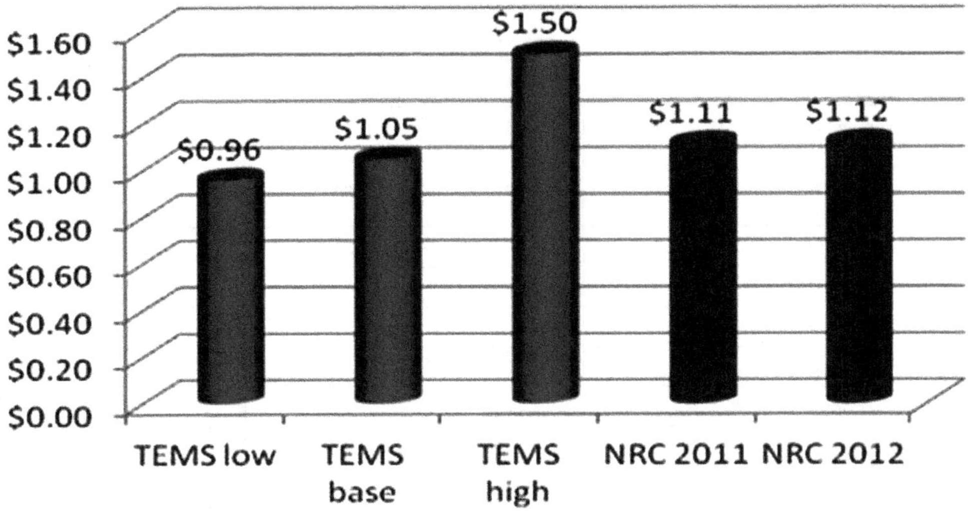

for TEMS high demographic ridership. Similarly, the advertising and commissions costs are also based on TEMS high demographic ridership in proportion to 2013. The Table 15 presents the updated annual operation costs.

According to Hurley and Jones (1990), once the infrastructure exists, HSR will have substantial cost advantage over other transportation modes. Further learning will contribute to lower operating costs over time. A French analysis by Berlioz and Leboeuf (1986) indicated that due to learning, operating costs would decrease over time (their study noted an average decrease of 48% over five years of operation). Hurley and Jones (1990) also note that this factor has not been included in Canadian HSR feasibility studies.

Financial and Economic Analyses

Financial and economic analyses are clearly two key factors that will influence the feasibility of HSR systems in Canada. These analyses will be based on the ridership, revenue and cost forecasts discussed previously.

It is noteworthy that in the feasibility report for the QWC HSR project, both financial and economic analyses covered eight scenarios, based on the two technologies F200+ and E300+ in each of the four main functional segments:

- Quebec City – Windsor, the base case scenario
- Quebec City – Toronto
- Montreal – Toronto
- Toronto – Windsor

Table 15. Estimated 2013 annual operating costs (in million 2013 CAD) (Van Horne Institute, 2013)

Cost Item	CPR	Greenfield Non-Electric	Greenfield Electric
Train maintenance	31.0	36.7	34.1
Track maintenance	3.2	21.5	25.7
Facility maintenance	4.4	5.2	5.1
Total maintenance costs	**38.6**	**63.4**	**64.9**
Fuel/electricity	15.7	16.8	5.5
Train crew	5.7	2.8	2.2
On-board service & product	10.7	14.4	18.7
Stations services	3.1	3.9	3.9
Commissary & call centre	2.3	3.3	4.5
Training	1.0	1.0	1.0
Insurance	2.9	5.1	5.1
Ads & commissions	8.3	8.3	16.8
Total operating costs	**49.5**	**55.6**	**57.6**
Admin & other	2.6	3.6	3.7
Total O & M costs	**90.8**	**122.6**	**126.2**
Contingency 2%	1.8	2.5	2.5
Total w/Contingency	**92.6**	**125.1**	**128.7**

Financial Analysis

The objectives of the financial analysis are mainly to (i) estimate the financial internal rate of return (IRR), (ii) calculate the level of required public funds, and (iii) define a mechanism for including private sector participation (PSP) in the funding of the project (EcoTrain, 2011).

To perform a financial analysis, the project structures or delivery modes need to be defined first. Therefore, in the up to date QWC HSR feasibility report by EcoTrain, the financial analyses was based on the representative project structures of the current market trends and approaches for HSR projects in other areas, and assumed that the project could be either entirely financed through public funds (wholly public case) or partly funded by the private sector (PSP case) over a 30-year commercial operation. In both cases, EcoTrain suggested setting up a High Speed Rail Authority responsible for the execution and operation of the project and entering into contracts with infrastructure providers chosen in the PSP case. Further, under the PSP case, the High Speed Rail Authority will be responsible for the HSR operation and assume the revenue risk.

To calculate the net present value (NPV) and the financial IRR for the QWC HSR project, a discount rate of 4.2%, based on the blended borrowing rates of the three governments of Canada, Quebec and Ontario was applied. Besides, it is assumed that the IRR required by the private sector is 13.5% based on the analysis of similar HSR projects worldwide. The main results of the financial analysis for both the wholly public case and the PSP case of the whole QWC HSR project are shown in Table 16.

From Table 16, it is seen that the NPV for the governments of the whole QWC HSR project will always be negative. The NPV will remain negative even when considering sales tax on fares returned to governments as revenue. The IRR for the governments in the PSP case will be negative for both technologies. As a result of the debt and equity financing provided by the private sector, the level of investment required from the governments in PSP can be reduced by up to 50%. However, since the private sector requires for higher rates of return, the financial return to governments would be less favorable under the PSP case than under the wholly public case. Given the available data and the assumptions determining the project's costs and revenues, the financial analysis results conclude that the present form of the QWC project is not sustainable. Also, a wholly private sector was not considered as an option since transferring all project risks to the private sector renders it impossible to finance.

Similarly, for the CEC HSR system, the financial analysis in the 2004 feasibility study also examined two project structure options to finance HSR construction and operation: an wholly publicly financed option and a shared public-private financing option over a 30- year operation period. The wholly privately financed option was not considered in the CEC HSR project either. The assumptions of the interest rates

Table 16. Main results of financial analysis (NPV in billion 2009 CAD) (EcoTrain, 2011)

Case for the QWC	Technology	Project NPV to Governments in 2009 billion CAD	Project IRR to Governments in Percentage	NPV of Investments by Governments in 2009 billion CAD
Wholly Public	F200+	-7.7	0.8	16.3
	E300+	-8.5	0.9	18.2
PSP	F200+	-14.7	-8.3	8.3
	E300+	-16.4	-7.5	9.2

used in the analysis are 5.2% for publicly funded debt, 7.2% for privately financed debt and 13% for return on equity.

In the financial analysis of the 2004 study, for the wholly publicly financed option, it was concluded that operation costs could be totally covered from year one. The debt and interest on the capital investment could be covered by sufficient surplus revenue generated in 25 years for the CPR option and 36 years for the Greenfield option. In the shared public-private financing option, operation costs are totally covered by the private sector. The surplus revenue generated could repay 34% to 66% of the total initial capital investment over 30 years.

Compared to the study in 2004, the interest rates and expected return on equity in the latest CEC HSR feasibility study are at least 2% lower. As a consequence, it is expected that savings in financing costs can either reduce the repayment period of the initial investment or lower required debt payment costs over the term of the investment. Further, it is more and more common that the debt repayment periods for public investments in fixed infrastructure are longer than 30 years so as to be consistent with their actual effective life, and to improve the affordability and potential attractiveness of these projects for private investors.

Economic Analysis

The objective of the economic analysis is to evaluate the feasibility of the project and its potential contribution to the economy as a whole in economic terms by considering non-financial costs and benefits such as changes in atmospheric emissions, public safety improvements, impact of HSR on transport operators within the corridor, terminal value, as well as the consumer surplus (EcoTrain, 2011). The method used in the economic analysis in the latest QWC HSR feasibility study by EcoTrain is the cost-benefit analysis which includes considerations such as consumer surplus, changes in atmospheric emissions, improvements in public safety and reduction of injuries and fatalities as well as impacts of the proposed HSR service on VIA Rail and other transport modes. The analysis results are shown in Table 17.

Based on the cost-benefit analysis results in Table 17, it is known that, from the perspective of the whole Canadian economy, there are three combinations of technologies and segments that have a positive NPV (in decreasing order of importance):

- E300+ on the Montreal – Toronto segment with NPV of $869 million CAD
- F200+ on the Montreal – Toronto segment with NPV of $817 million CAD
- E300+ on the Quebec City – Toronto segment with NPV of $257 million CAD

Table 17. Main results of the economic analysis (NPV in million 2009 CAD) (EcoTrain, 2011)

Segment		F200+	E300+
Quebec City - Windsor	Revenues	9536	10677
	Consumer surplus	1741	2665
	Atmospheric emissions	47	96
	Public safety	2363	2481
	Impact on VIA Rail	1175	1175
	Terminal value	581	650
	Investment costs	-11291	-12622
	Operating costs	-4091	-4284
	Renewal costs	-1175	-1539
	Impact on other modes	-253	-293
	Total	**-1376**	**-992**
Quebec City – Toronto		-190	257
Montreal – Toronto		817	869
Toronto - Windsor		-1039	-1433

Allocating the cost and benefit provincially to Ontario and Québec, the results of the economic analysis illustrated in Table 18 show that, from the point of view of the Ontario economy as a whole, all scenarios are economically feasible as they all have positive NPVs, except the combination of E300+ technology and Toronto – Windsor segment which has a negative NPV. Whereas regarding the Quebec economy as a whole, there are only three scenarios that have positive NPVs which are presented to be economically feasible (in decreasing order of importance):

- F200+ on the Montreal – Toronto segment with NPV of $116 million CAD
- E300+ on the Quebec City – Toronto segment with NPV of $108 million CAD
- E300+ on the Montreal – Toronto segment with NPV of $80 million CAD

Although in a different order, the three economically feasible scenarios from the point of view of the Quebec economy as a whole are the same as those from the point of view of the Canadian economy. There is no individual economic analysis for the CEC HSR system in the feasibility studies by Van Horne Institute.

Table 18. Main results of the economic analysis from the point of view of the Quebec and Ontario economies (NPV in million 2009 CAD) (EcoTrain, 2011)

Segment		F200+		F300+	
		Ontario	Quebec	Ontario	Quebec
Quebec City - Windsor	Net contribution by governments	-3321	-2396	-3673	-2651
	Consumer surplus	1104	637	1673	991
	Atmospheric emissions	36	11	63	34
	Public safety	1618	735	1708	774
	Impact on VIA Rail	524	215	524	215
	Impact on airlines	-40	-14	-45	-17
	Impact on airports	-153	-53	-174	-64
	Impact on bus operators	6	2	5	2
	Terminal value	275	198	308	222
	Total	**49**	**-665**	**388**	**-494**
Quebec City - Toronto		362	-132	661	108
Montreal - Toronto		779	116	983	80
Toronto - Windsor		62	-691	-49	-878

Impacts on the Transportation System

The impacts on the transportation system due to the introduction of the HSR system can be generally classified into direct impacts and the indirect impacts. The potential direct impacts on competing modes are best evaluated by the modal shifts from existing modes of travel to HSR. Trips induced by HSR (trips by HSR that will not be made in the absence of HSR services) are also a kind of direct impact.

As an efficient transport mode, from the perspective of travel time, HSR is very competitive. From the feasibility report by EcoTrain for the QWC HSR, Table 19 illustrates the travel times calculated between all stations for major city pairs. From this table, it is seen that the introduction of HSR in QWC will significantly cut down the travel times between major cities pairs. Moreover, the downtown to downtown travel times between cities by HSR can even be competitive with air services.

In the 2004 CEC HSR feasibility study by Van Horne Institute, the travel time savings were estimated by calculating door-to-door travel time using average access and egress time for all modes and in-terminal and waiting time for air, bus and HSR in addition to en-route travel time. The average door-to-door travel times estimated for each mode are shown in the following Table 20.

Table 19. Travel times for major city pairs in QWC (EcoTrain, 2011)

City Pair	Existing VIA Rail Service[a]	F200+	E300+
Quebec City – Montreal	3 hr 09 min	1 hr 49 min	1 hr 26 min
Montreal – Ottawa	1 hr 55 min	1 hr 11 min	0 hr 57 min
Ottawa – Toronto	4 hr 36 min	2 hr 25 min	1 hr 50 min
Montreal – Toronto	5 hr 12 min	3 hr 38 min	2 hr 47 min
Toronto – London	2 hr 07 min	1 hr 05 min	0 hr 51 min
Toronto - Windsor	3 hr 59 min	2 hr 02 min	1 hr 33 min

[a]Running times based on 2010 VIA Rail schedules are average for the number of trains per day on each route.

Table 20. Estimated door-to-door travel times (minutes) (EcoTrain, 2011)

Calgary – Edmonton	Car	Air[a]	Bus	CPR HSR	Greenfield Electric HSR
In-Vehicle	180	55	200	130	97
Access/Egress	15	75	30	30	30
In-Terminal	0	65	15	15	15
Total	**195**	**195**	**245**	**175**	**142**
Red Deer					
In-Vehicle	90	N/A	90	65	48
Access/Egress	15	N/A	15	15	15
In-Terminal	0	N/A	15	15	15
Total	**105**	**N/A**	**120**	**95**	**78**

[a]Assumes flights between Calgary International and Edmonton International airports.

The results in Table 20 indicate that the CPR option can offer a 20-minute saving in travel time compared with both car and air travel and a 70-minute travel time saving compared to bus travel between Calgary and Edmonton. With the Greenfield Electric option, the reduction of travel time is expected to increase to 53 minutes relative to car and air travel, and 103 minutes compared with bus travel. As HSR users are more likely to include individuals whose origin and/or destination are close to the stations, and the future in-vehicle travel time required for both car and bus are likely to increase due to growing congestion, these forecasts are most likely conservative. Based on the assumptions about business and non-business travelers' value of time as well as average wage for adults aged 25-54, the NPV of travel time savings associated with the CPR option and the Greenfield Electric are $317 million and $332 million respectively over the 30-year life of the project.

For both business and non-business travel, travel cost savings represent the net difference between the current cost of travel by other modes and the cost of HSR. With assumptions on the proportion of total household budget on transportation, car occupancy, perceived average vehicle operating cost per km, the NPVs of travel cost savings over the 30-year life of the project associated with the CPR option and the Greenfield Electric option are $603 million - $1,128 million CAD and $756 million - $1,258 million CAD respectively depending on assumed car occupancy.

Figure 10 compares the estimated full cost per passenger-km for different modes as reported by Hurley and Jones (1990). This cost advantages along the reduction in travel times due to the introduction of HSR systems are significant in both QWC and CEC areas. Hence, it is very likely to have influences on existing transportation services operating within the corridors namely intercity bus, air, automobile, commuter rail and VIA Rail services. These impacts can be best measured by the modal shifts from existing modes of travel to HSR. Besides, freight rail services may also be influenced by HSR where routes are planned to share the usage of corridor RoW with freight railways.

In the study by EcoTrain for the QWC HSR system, to estimate the modal shifts, the total number of trips by existing transport modes was determined for the 2006 base year and growth was estimated for 2031 and 2041 for both options F200+ and E300+. Induced ridership to HSR was also forecasted. The forecasting results indicate that the total HSR ridership in 2031 for F200+ option and E300+ option are 10.23 million and 11.13 million respectively. The projected percentage of HSR ridership shifted from other modes is 93% to 94% which mainly originates from the

Figure 10. Estimated cost per passenger-km by mode (Hurley, W.J., & Jones, J. 1990)

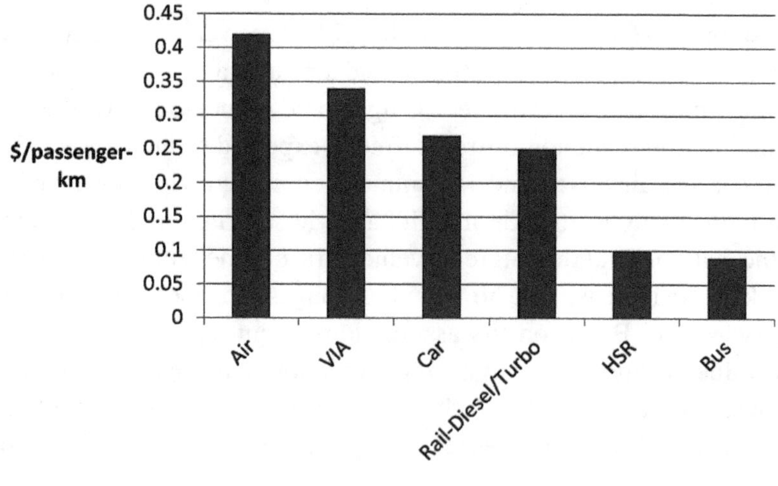

auto mode with a proportion of 59%, and only 6% to 7% of the ridership is induced by the availability of HSR. Besides, the forecasts in 2041 are similar to those in 2031.

Other potential direct and indirect impacts on the transport system resulting from the introduction of HSR service in QWC are summarized as follows:

- It is estimated that 6.1 million of the total 83 million auto trips per year will transfer to F200+ HSR in 2031. The implementation of the HSR in QWC is not expected to result in significant savings in road infrastructure requirements.
- Comparing with the forecasts in the QOHSRPS, the air market loss in the QWC in 2031 estimated by EcoTrain will be lower, at 31% for the F200+ option and 36% for the E300+ option. Also, the airport improvement fees assessed to the users of various airports in QWC will be directly influenced by the expected ridership diverted from air to HSR.
- For rail transport, the study by EcoTrain has estimated that 70% of VIA Rail ridership will transfer to HSR in 2031 which is 10% higher than the forecast in the QOHSRPS. In addition, in highly urbanized areas, HSR may have to share tracks with commuter trains and freight trains which would require a high level of coordination and a modernized signal and telecommunication system to maximize capacity and service performance.
- Compared to other transport modes, only the estimated percentage of ridership diverted from bus to HSR is negative (i.e., there will be a shift to cheaper bus travel as a result of HSR replacing conventional rail), between -3% and -4%. It is also expected that eliminating local station stops served by VIA Rail previously will push some travelers transferring from rail to bus.
- Moreover, the analysis of the impacts on other transportation modes also shows that the integration and interconnectivity of HSR with other transportation modes (both existing and planned) are essential to make HSR a success in the QWC.

For the analysis of the impacts on the transportation system by the CEC HSR project, a modeling approach based on trip diversion factors was used in the feasibility study in 2004 to forecast the overall market responsiveness to the HSR service, and the importance of trip time, frequency and price in attracting travelers from other transport modes. The analysis was based on both the low and average total number of trips of the 2003 market research and the CEC HSR service plan. Assuming a mature HSR operation in 2003, ridership forecasts were developed for each of the three representative technology options (CPR, Greenfield Non-Electric and Greenfield Electric). The midpoint in the range of ridership forecasts are presented in Table 21.

Table 21. Ridership forecast base year – 2003 – based on midpoint low/avg (Van Horne Institute, 2004)

Person Trips ('000)	Car	Air	Bus	HSR
CPR	-947	-189	-184	1320
Greenfield non-electric	-1148	-208	-184	1540
Greenfield electric	-1250	-225	-193	1668

From the analysis results, it is estimated that HSR ridership will range between 1.3 and 1.6 million trips depending on the representative technology option selected which represents a 22%-28% share of the total travel market. Nearly three-quarters of forecast HSR riders are diverted from current auto users. However, this is a relatively small proportion of the total auto trips and leaves auto as the dominant travel mode with a 68%-73% market share of the total trips. The rest of HSR riders are current air and bus travelers, particularly the business travelers due to HSR's high service frequency and travel time reduction relative to the other transport modes.

To forecast the induced trips by CEC HSR service, by applying suitable discount factors to only those potential travelers that stated they would be very likely to use HSR, and assuming each will take one new round trip per year because of the presence of HSR (induced demand), the ridership forecast predicts increases of 4% to 5%.

Socio-Economic and Environmental Impacts and Benefits

Previous sections discussed HSR concepts and fundamentals regarding the implementation of HSR in Canada. However, given the huge investments required to implement HSR, a discussion of the social, economic and environmental impacts is important.

Before analyzing the environmental and social impacts, it should be noted that many changes have occurred in the environmental evaluation and assessment process in Canada during the past twenty years. As Canadians' perceptions of environmental issues and sustainability are becoming prominent, there is more incentive to ensure that environmental issues influence the design and implementation of the project. Also a greater level of participation and consultation with stakeholders during project planning is necessary.

To evaluate the environmental impacts caused by the QWC HSR project, besides inventorying and mapping natural and social environmental issues in the QWC and identifying interactions between these issues and the HSR representative routes on a segment-by-segment basis, EcoTrain also analyzed safety, noise and vibration and atmospheric emissions issues. However, it was not possible to identify all environmentally sensitive locations and issues. The selected eleven environmental issues include six natural environmental issues, and five social environmental issues, namely (EcoTrain, 2011):

Natural Environment:

- Substantial Watercourses (includes rivers, streams, brooks, or creeks)
- Wetlands
- Natural Protected Areas
- Sensitive Fish Habitats
- Prime Agricultural Lands (Categories 1 and 2 soils as defined by Natural Resources Canada)
- Resource Extraction Areas

Social Environment:

- Residential Areas
- Recreation and Tourist Areas
- Historical or Cultural Site and Cemeteries
- Federal Reserves, including First Nation reserves, Land Claim Settlement Lands, and military bases
- Waste Management Sites

On a segment-by-segment basis, the interactions between the issues proposed above and the HSR representative routes are identified. Quantifying the impacts and estimating the associated costs are not possible for all environmental and social issues except for safety, noise and vibration, as well as atmospheric emissions.

For safety issues, the fatalities and injuries data for air, rail and road in Ontario and Quebec were from Transportation Safety Board of Canada (TSB), Transport Canada (TC), Ontario Ministry of Transportation (MTO) and Société de l'Assurance Automobile du Quebec (SAAQ). The expected benefit obtained from preventing a fatality or injury was measured based on the value of a statistical life (VSL), $4.96 million CAD (2009) for fatalities base value and $318,000 CAD (2009) for injuries base value.

The total number of fatalities and injuries in 2031, with and without QWC HSR, as well as the total reduction in the associated economic losses (in 2009 CAD) and the percentage reduction from the existing (2005) conditions are summarized in Table 22.

Table 22. Reduction in fatalities and injuries due to HSR, 2031(Van Horne Institute, 2013)

	Ontario						Quebec					
Total	**Without HSR**		**With F200+**		**With E300+**		**Without HSR**		**With F200+**		**With E300+**	
	Number	**Value***	**Number**	**Value**	**Number**	**Value**	**Number**	**Value**	**Number**	**Value**	**Number**	**Value**
Fatalities	219.6	$1090		$186		$196	89.1	$442		$80		$64
Serious Injuries	445.2	$142		$20		$21	331.8	$106		$14		$83
Economic	Loss: $1232 Loss Avoided: $206						Loss: $548 Loss Avoided: $93					

*Values are in million 2009 CAD

On the basis of the similar projects recently built in the QWC, in the cost analysis by EcoTrain, mitigation costs for noise barriers, settlement ponds, farm crossings and wildlife passages, and included other mitigation costs either in the costs of earthworks, drainage and structural works were explicitly calculated (EcoTrain, 2011).

For atmospheric emissions issues, the total economic benefit from the reduction in pollutants by the implementation of F200+ option and E300+ option in QWC were estimated at approximately $192 million CAD (2009) (with a NPV of $47.4 million CAD) and $390 million CAD (2009) (with a NPV of $96.1 million CAD) respectively.

Additional natural and social environmental issues which cannot be mapped at the scale of this QWC HSR project are also identified and listed in the study by EcoTrain. Particularly, these additional impacts include the social-economic effects on residents living close to the HSR lines or stations caused by the implementation of HSR namely increases or decreases in property value, property displacement and acquisition, job creation and loss, loss or gain of connectivity and access, nuisance effects (noise, vibration and dust) during construction and operation as well as visual effects (EcoTrain, 2011).

The 2004 Van Horne Institute report had a detailed discussion on the socio-economic impacts and benefits in the Alberta context as summarized in Tables 23 and 24.

Of the bottom line benefits, the one related to the user is most important. In the case of HSR, it is reduced travel time and cost. Other benefits include less stress for car drivers, better comfort, more productive time, and improved travel reliabil-

Table 23. CEC HSR socio-economic and environmental benefits: quantifiable (Van Horne Institute, 2004)

Summary of Quantifiable Benefits		
	CPR	**Greenfield Electric**
User Benefits: • Travel time savings ($M NPV) • Travel cost savings ($M NPV)	 $317 $603 – 1,319	 $332 $756 – 1,258
Economic Development: • Employment expansion (# jobs) • Employment Income ($M NPV)	 1,600 jobs $619	 2,600 jobs $922
Social Benefits: Accident reduction ($M NPV)	 $215	 $72
Environmental Benefits[a]: Reduction in GHG value ($M NPV)	1.8 million metric tonnes $56	3.1 million metric tonnes $98
Financial Benefits[b]: • Construction employment • Const. employ. income ($M NPV) • Operations employment (# jobs) • Ops. employ. income ($M NPV) • Tax revenues - Alberta • Tax revenues - Canada	 25,500 person-years $983 1,000 -1,100 jobs $491 $172 - $343 $378 - $707	 52,000 person-years $1,950 1,350 -1,450 jobs $861 $365 - $565 $800 - $1,185

[a]Greenfield Electric reduction is based on direct emissions only; adjustment for emissions from power generation would reduce this reduction and its associated value.

[b]Includes all direct, indirect and induced employment, employment income and taxes on this income

Table 24. CEC HSR socio-economic and environmental benefits: non-quantifiable (Van Horne Institute, 2004)

Summary of Non-Quantifiable Benefits	
Economic Development: Competitive image	Positive – catalyst for economic growth. Potential for corridor to be perceived internationally as a single economic unit. Opportunity for greater economic cooperation and development between communities in the corridor. Increases development in Red Deer and near suburban stations.
Transportation Network Optimization: • Road congestion & highway infrastructure • Inter-city bus transportation • Air transportation • Rail freight operations & shippers	 Could ease congestion particularly at chokepoints. May result in deferral of infrastructure. May result in reduced service and job losses unless mitigation applied. Likely to spur specialization, increased choice & better service; no deferral of infrastructure or job losses expected. Reduced freight delays and improved speeds; potential for new intermodal freight service; and, enables upgrading of line south of Calgary using surplus rail materials. May result in shipping costs savings, expanded business opportunities and/or new business potential for Alberta industries.

continued on following page

Table 24. Continued

Summary of Non-Quantifiable Benefits	
Social Benefits: Re-shaping of growth & development	Shifts and focuses development to central area of corridor and relieves pressure on land and prices in Calgary and Edmonton but requires planning and controls to avoid urban sprawl.
Environmental Benefits: Noise	CPR: Some reduction in highway noise but may increase noise along rail corridor. Greenfield: Some reduction in highway noise and less noisy than CPR alternative due to electric propulsion and dedicated portion of corridor away from settlements.

ity, not insignificant benefits given the severe winters in Alberta. Further the traveler now has more choice, road congestion would improve and the government could probably spend less on road transportation infrastructure. An important social benefit would be accident reduction on the Calgary-Edmonton highway and the resultant saving of lives and injuries.

Another significant benefit was the contribution of the new HSR to economic development. Since HSR would effectively shrink the distance between Calgary, Red Deer and Edmonton, it was seen as integrating the economies of these three regions improving the access to labor and markets (both customer and supplier). The whole region would become more attractive, thus inducing more business investment and labor from outside Alberta, leading to strong economic development. It was felt that Red Deer would relatively benefit more but this would not be a problem for Edmonton or Calgary as it would be a net gain from outside the three regions. All three regions would show strong growth.

Currently Calgary and Edmonton are seen as 'complementary' cities with Calgary hosting relatively more corporate offices and Edmonton being the operations center. However with growth in the cities, each is seen as becoming more independent. HSR is seen as proving a strong fillip to this trend establishing both as vibrant diverse economies. By being seen as an integrated economic region, Alberta will be much more globally attractive. Some positive effect on property values is also seen.

Clearly, HSR would have a negative effect on the bus and air transportation sectors. The 2004 report speculated that while there would indeed be a negative effect on the bus transportation industry (unlike the QWC corridor where HSR would replace conventional rail, which would result in some travelers shifting to bus as explained earlier, this corridor has no rail currently) it could be offset if the companies focused on other markets. Indeed currently there is great need for bus transportation in the oil sands industry. The oil sands are remote and spread over a large area. Many large international oil and gas companies have projects in these remote areas and have constructed temporary housing for staff off site. Thus, staff has to be transported daily from their residences to the site and within the sites. This is an example of a new market that bus transportation companies can focus on.

With respect to air transport, about one half of the 500,000 passengers travelling between Edmonton and Calgary connect to flights outside the region. Naturally they would not be affected. Thus the HSR effect is not seen to be significantly negative. In any case, since the 2004 report, with the strong growth that Alberta has enjoyed in recent years both airports have experienced growth and the 2004 prediction of no job losses has been borne out. Naturally, there would be a lot of temporary jobs created during construction.

Particularly, if the CPR option is exercised, there would like be improvements in freight rail operations because of the improved rail infrastructure. Such a CPR short haul high performance intermodal service already exists in Central Canada and might become feasible in the CEC corridor. CPR suggests that there might be track material made surplus by this project that could be redeployed to conventional rail freight transport infrastructure, improving connections to the USA border and British Columbia. Overall there are many quantifiable and non-quantifiable economic benefits that may be realized. This of course would also benefit government revenues.

Greenhouse gas (GHG) emissions would decrease because of less automobile and air travel but as mentioned earlier, currently Alberta produces much of its electricity from coal so HSR would not be as environmentally friendly as in the QWC corridor. However with Alberta's electricity generation expected to move more to natural gas, the situation should improve. Noise pollution is not seen as an issue.

CONCLUSION

This chapter discussed the feasibility of two potential HSR projects in Canada based on the latest and previous studies. It also generally reviewed the existing HSR systems worldwide and lessons that may be applicable for Canada. A variety of socio-economic and environmental benefits that can be brought by the introduction of the QWC and CEC HSR projects have been shown.

From the probability of construction perspective it seems to be good news and bad news at the same time. The good news is that it appears the Ontario Government is moving forward with a section of the QWC. The government is planning to do this on its own rather than as a public private partnership. It is hoped that the expected success of this project would provide motivation to expand HSR on other sections of the QWC. Logically, it makes sense that this is the region that would move forward first as it is the most densely populated corridor in Canada and most likely to succeed in terms of ridership. The prognosis for the other corridor, the CEC does not look promising in the short term though there is recognition that

HSR would be an option sometime in the future. A recent report commissioned by the Alberta Legislature recommended not proceeding with HSR currently but still recommends that the government create a regulatory framework for HSR, as well as a utility corridor that would include space for HSR.

In addition, it is expected that the analysis approaches used and other lessons that can be learned in the feasibility studies of the two HSR projects in Canada can also be applied to other jurisdictions that may have similar conditions.

REFERENCES

Albalate, D., & Bel, G. (2012). High-speed rail: Lessons for policy makers from experiences abroad. *Public Administration Review*, *72*(3), 336–349. doi:10.1111/j.1540-6210.2011.02492.x

Alberta Association of Municipal Districts and Counties. (2010, October 1). *Study of high speed rail impacts on rural Alberta, final report*. Retrieved from http://www.aamdc.com/component/docman/doc_details/78-2010-study-of-high-speed-rail-impacts-on-rural-alberta?Itemid=208

Balakrishnan, J., Eliasson, J. B., & Sweet, T. R. C. (2007). Factors affecting the evolution of manufacturing in Canada: An historical perspective. *Journal of Operations Management*, *25*(2), 260–283. doi:10.1016/j.jom.2006.05.014

Berlioz, C., & Leboeuf, M. (1986). Performance of the Paris-southeast TGV. *Revue General des Chemins de Fer*, *196*, 759–768.

Bernard, A. B., Moxnes, A., & Saito, Y. U. (2014, September 24). Fast trains, supply networks, and firm performance. Centre for Economic Policy Research, UK. Retrieved from http://www.voxeu.org/article/fast-trains-and-firm-performance-new-japanese-evidence

Canadian Broadcasting Corporation. (2014, April 15). Liberals promise high speed rail in Kitchener within 10 years. Rail line would link Toronto, Pearson, Kitchener and London. Retrieved from http://www.cbc.ca/news/canada/kitchener-waterloo/liberals-promise-high-speed-rail-in-kitchener-within-10-years-1.2610303

Craven, P., & Traves, T. (1983). Canadian railways as manufacturers, 1850-1880. *Historical Papers*, *18*(1), 254–281.

de Rus, G. (Ed.), (2009). *Economic analysis of high speed rail in Europe*. Bilbao Spain: Fundación BBVA. Retrieved from http://www.fbbva.es/TLFU/dat/inf_web_economic_analysis.pdf

Dykstra, M. (2014, May 23). Edmonton and Calgary should think about high-speed rail system, but shouldn't put money into it. *Edmonton Sun*. Retrieved from http://www.edmontonsun.com/2014/05/23/edmonton-and-calgary-should-think-about-high-speed-rail-system-but-shouldnt-put-money-into-it

EcoTrain. (2011). *Updated feasibility study of a high speed rail service in the Quebec City – Windsor corridor*. Montreal, Quebec: EcoTrain.

Government of Alberta. (2013, September 26), *Alberta's population reaches four million*. Retrieved from http://alberta.ca/release.cfm?xID=350618FCB7CB9-D315-C7B6-3A1E8521B366D28A

Government of Alberta. (2014, September 30). Electricity facts. Retrieved from http://www.energy.alberta.ca/Electricity/681.asp

Greenlaw, C. C. N. (2007). *VIA rail*. St. Paul, MN: Voyageur Press.

Hurley, W. J., & Jones, J. (1990). *The feasibility of high-speed rail in Canada*. Kingston, Ontario: Department of Economics, Queen's University.

International Union of Railways. (2011, July 27). UIC world statistics. Retrieved from http://www.uic.org/com/IMG/pdf/cp18_uic_stats_2010_en.pdf

International Union of Railways. (2014, September 1). High speed lines in the world. Retrieved from http://www.uic.org/IMG/pdf/20140901_high_speed_lines_in_the_world.pdf

Jeanes, D. (2009, May 28). *Meeting of the Standing Committee on Transport, Infrastructure and Communities*. Parliament of Canada, Ottawa, Ontario. Retrieved from www.parl.gc.ca/HousePublications/Publication.aspx?DocId=3935511&Language=E&Mode=2&Parl=40&Ses=2

Central, J. R. (2015, February 1). Retrieved from http://english.jr-central.co.jp/company/company/achievement/financeandtransportation/transportation2.html

Pepy, G. & Perren, B. (2006). 25 years of the TGV. *Modern Railways*, October, 67-74.

Pomfret, R. (1981). *The economic development of Canada*. Toronto, Ontario: Methuen.

Smith, K. (2014, May 2). Ontario government backs Toronto - London HSL. *International Railway Journal*. Retrieved from http://www.railjournal.com/index.php/high-speed/ontario-government-backs-toronto-london-hsl.html?channel=542

Standing Committee on Alberta's Economic Future of the Legislative Assembly of Alberta. Edmonton. (2014, May). *The feasibility of establishing a high-speed rail transit system in Alberta.* Retrieved from http://s3.documentcloud.org/documents/1173385/high-speed-rail-transit-system-in-alberta-final.pdf

Transport Canada. (2014). Retrieved from http://www.tc.gc.ca/eng/rail-menu.htm

Van Horne Institute. (2004). *Calgary/Edmonton high speed rail: an integrated economic region.* Calgary, Alberta. The Van: Horne Institute.

Van Horne Institute. (2013). *Updated cost and ridership/revenue for Calgary/ Edmonton high speed rail.* Calgary, Alberta. The Van: Horne Institute.

World Bank. (2014, July 10). Retrieved from http://www.worldbank.org/en/news/press-release/2014/07/10/cost-of-high-speed-rail-in-china-one-third-lower-than-in-other-countries

World Market Intelligence News. (2014, July 15). *Ministry of transport - London to Toronto high-speed Rail line – Ontario.* Retrieved from http://ezproxy.lib.ucalgary.ca/login?url=http://search.proquest.com/docview/1552723667?accountid=9838

KEY TERMS AND DEFINITIONS

Demographics: Examining the dynamics of population such as growth, movement employment, culture, and income.

Economic Analysis: The objective of the economic analysis is to evaluate the feasibility of the project and its potential contribution to the economy as a whole in economic terms by considering non-financial costs and benefits.

Emissions: Emitting of chemicals that cause global warming as well as atmospheric pollutants.

Greenfield Rail: Building a completely new railway line as compared to using existing railway track.

High Speed Rail: Passenger rail transport that typically can reach speeds of over 125 miles/hr. (200 km/hr.) and up to 220 miles/hr. (350 km/hr.).

Net Present Value (NPV): In calculating the profitability of a project, discounting future revenues and costs to determine the profit in today's currency value.

Socio-Economic Analysis: Examining not only the costs and benefits of an endeavor based on economics but also examining the social impact including on the environment, some of which may not be quantifiable.

Standing Committee: A permanent committee in an organization that is tasked with looking at relevant issues related to focused topics.

Subsidy: A financial contribution (often by governments) provided to facilitate any activity that would not occur without this contribution.

Chapter 10
High Speed Rail:
Study, Report, Current, and Future Considerations

Raj Selladurai
Indiana University Northwest, USA

George VandeWerken
Providence Bank, USA

ABSTRACT

This chapter focuses on the recent Southwest states' multi-state planning collaborative study, which the Federal Railroad Administration initiated, coordinated, and released in 2015. It highlights the process, planning contextual factors, conclusions, recommendations, and implications, which may serve as a reference model for future studies in other regions of the U.S. The chapter also looks at the momentum of high-speed rail in the U.S., especially in California, and explores some of the considerations in moving toward more optimal and comprehensive transportation planning and policy development for high-speed rail for the U.S. in the future.

INTRODUCTION

High-speed rail is gaining momentum in many countries, including the United States -- especially in the states of California, Florida, Texas, and other travel corridors. Dumaine (2015), in his report, has stated that plans for bullet trains in the United States are "gathering steam" and "projecting" superfast forward toward potential

DOI: 10.4018/978-1-5225-0102-2.ch010

implementation. The study cited several emerging factors such as growing populations, crowded, congested highways, and the new generation's apathy toward cars and car ownership that have stimulated and encouraged a new passion and interest in high-speed rail, especially in some southern states. As a further indication of an apparent changing national sentiment, the U.S federal government has recently approved a $28 million grant for a Maryland feasibility study to develop an unprecedented, superfast 300 mph Maglev train from Washington D.C. to New York City. This train would complete the trip that currently takes at least three hours in less than one hour (Kelly, 2015).

Kanter (2015), a Harvard University scholar, observed that Japan and other countries like Germany, France, and China have invested heavily in high-speed rail and new, state-of-the-art infrastructure for passenger rail systems, whereas the U.S. has chosen to focus on highways and still relies on 19th century rail infrastructure. This has resulted in some major problems related to safety, congestion, and service failures in the US, which have affected our economy, health, and overall quality of life. Kanter further recommended that passenger rail in the US should emulate its counterpart in the freight transportation rail area where the US is considered a model leader for the rest of the world. Private companies such as CSX can invest in new infrastructure and other related developments due to their significant growth in the freight area. Perhaps we need a blended approach of private and public funding/support of passenger high-speed rail to keep up with the similar blended efforts being carried out successfully in the freight area.

Another recent report has focused on the fast development of a rail service in Toronto, Canada called the UP Express, a premium airport express train linking Toronto Union Station to Toronto Pearson International Airport (Cross Rail Chicago, 2015). Interestingly, Mayor Rahm Emmanuel and his Chicago leadership are looking closely at this Canadian rail model while considering similar rail service proposals connecting Metra to a station near O'Hare airport in Chicago.

Further, the Federal Railroad Administration (FRA) has recently released its "Corridor Alternatives Analysis Technical Report" for the Dallas to Houston high-speed rail project. Proposed by the Texas Central Railway (TCR), this report evaluates potential corridors, stations, and different types of related services. The proposed project would span a 240-mile corridor, on dedicated high-speed rail track, which means it would be used for passenger rail service. The route would be built and operated by TCR, a private company, and would be able to move passengers between the cities of Dallas and Houston in 90 minutes by 2021 (Federal Railroad Administration, 2015).

Several timely reports and findings have highlighted the benefits of high-speed rail, especially as projected for California, and have further enhanced the momentum for high-speed rail. In a recent focus group study conducted at a local university campus, Selladurai (2013) reported that the business leaders perceived high-speed rail to provide many positive benefits, including saving time, safer travel, convenience, reducing highway congestion, and protecting the environment. Another report discussed the strong, positive benefits of high-speed rail to the state of California, its residents, and economy, etc. similar to those in Europe (Kelly and Nichols, 2014). It contended that high-speed rail decreased air trips between London and Paris by 50 percent; helped about 60 percent of air travelers from Barcelona to Madrid to switch to rail; and enabled Los Angeles commuters near the Metro Expo Line to triple their public transit use and cut back their driving by 40 percent after rail was introduced. Similarly, in California, the writers projected that in about 25 years, high-speed rail would reduce driving by about 3.6 billion miles a year, the equivalent of taking 317,000 vehicles off the road daily. In addition, high-speed rail would reduce between 300,000 and 675,000 metric tons of greenhouse gas emissions because high-speed rail would operate with net-zero emissions and fully 100 percent renewable energy. The high-speed rail system would also lead to enhanced land use planning and economic development in the travel corridors like we already see in the city of Fresno and other related areas.

Finally, it is important to note that the current federal administration seems to have shown strong support for several high-speed rail related proposals promoting high-speed rail in the US. A recent Federal government-initiated multi-state rail planning collaborative study (SW Study) was conducted with six Southwest states that participated: California, Arizona, Nevada, Colorado, New Mexico, and Utah (Federal Railroad Administration, 2014). The study focused on these specific states as they historically have showed relatively more interest in the rail service activities.

HIGHLIGHTS OF THE SW STUDY

In the SW Study (Federal Railroad Administration, 2014), representatives from various transportation related agencies from the six states of California, Arizona, Nevada, Colorado, New Mexico, and Utah worked together to develop multi-state rail plans and a common long-term vision for High Performance Rail (HPR) networks. This effort provided an analytical framework for future HPR studies conducted in other states to consider and possibly follow.

The Study focused on two major objectives: (1) to identify, evaluate, and plan a multi-state network of corridors; and (2) to identify the challenges and opportunities of rail development and implementation. Several executive leaders and staff from multiple stakeholder groups including Amtrak, Arizona Department of Transportation, BNSF Railway Company, California High Speed Rail Authority, Union Pacific Railroad, Nevada Department of Transportation, San Diego Association of

Governments, and others participated in this study. The study focused on several contextual factors, including land development patterns that encompassed more than 379,000 square miles, including the three megaregions of Northern California, Southern California, and the Arizona Sun Corridor. With corridors of 100 to 600 miles, this megaregion area would be very appropriate and favorable for high-speed rail. HPR would strengthen connectivity between cities, reduce dramatically the travel times, help improve economic development in these areas, and create a huge network linkage of one "supermegaregion."

In terms of the next factor, demographics, the SW Study area focused on some of the highest population growth areas in the U.S. Nevada and Arizona projected the highest growth rates of 35 and 25 percent during 2000-2010. California continues to lead the nation with the highest population in the nation, and the 3.4 million new residents who moved into the state during 2000-2010 represented about 12.4 percent of the country's total population growth. Further, California's projections indicate that the state would experience high growth rates through 2050. Also, this SW area showed a high population concentration (94 percent) in urbanized areas, but these urbanized areas constituted only three percent of the total land area, indicating high population density. High-speed rail is most efficient when serving densely populated areas with high economic activity.

Then, in terms of economic activity, the SW area generates a combined gross domestic product of $2.3 trillion, which constitutes about 16 percent of the total U.S. economy. An interesting, important observation is that this area is also the largest economy in the world that does not yet have dedicated high-speed rail.

The strong projections for population growth and economic growth development clearly point to more growth in travel via all modes in this SW area. Forecasts have emerged showing about 70 percent growth in travel with annual trips increasing from 162 to 273 million. Although car travel between cities would perhaps remain the dominant mode, high-speed rail and other travel modes would potentially see significant increases too. Studies show that intercity travel from Los Angeles to other major cities like San Francisco, San Diego, Las Vegas, Phoenix, and Salt Lake City is projected to increase substantially in the future, and high-speed rail would be an attractive option to consider (Federal Railroad Administration, 2014).

Two key recommendations resulted from this study. The first one is to form a voluntary California-Arizona-Nevada Passenger Rail Policy and Planning Group to continue implementing a broader strategy to engage all stakeholders in developing and refining the current vision. The second one would be to organize a Blue Ribbon Commission to guide a feasibility study for the Phoenix-Southern California corridor.

Further, the SW study developed two major implementation strategies. To continue the momentum in high-speed rail planning for the Southwest, this study would be integrated into each state's existing and ongoing transportation plans and policies.

Secondly, the states should plan on setting up a Southwest Rail Working group to implement the study's recommendations.

This report provided some lessons for high-speed rail planning and policy for other travel corridors. These include stakeholders' involvement in determining plans, policies, and outcomes; integration of all modes of travel in the planning equation; recognition of the federal government's role in multi-state planning; development of goals, purpose, and need for HPR early in the planning process; and flexibility based on the specific region's needs in determining HPR policies.

FUTURE CONSIDERATIONS

Not only passenger service but also freight transportation innovation has been progressing globally. The United States has been generally recognized – even envied – as a leader in freight innovation and freight logistics expertise. The intermodal use of containers to move freight from water to rail to truck and back again is a great logistics success story, particularly with the high level of rail and trucking integration. Privately owned rail and trucking companies continue to improve the speed and movement of containerized shipments as well as bulk commodities. Computerized inventory systems that automatically control ordering and direct freight movements now allow for advanced inventory and supply chain management and just-in-time delivery.

On the other hand, innovation in the passenger movement business has been much slower to progress in the United States than in Europe, Japan, and Asia. Although U.S. airline-passenger traffic growth has been stellar, particularly during the second half of the 20th century, rail adoption and expansion – particularly high-speed rail – by contrast, has been very slow. As urban centers in the United States have developed, most urban planning has centered on building more airport capacity – and especially more roads for automobile and truck use. Light rail transit in most urban centers is frequently rather limited – often only the skeletal remains of systems designed and built in the early part of the 20th century. Commuter bus systems struggle to expand, and they clamor for the necessary public subsidies to keep them running. Moreover, the nation's principal intercity passenger rail system, Amtrak, faces ongoing funding resistance in Washington.

Several studies and reports have shown the lack of support for high-speed rail in this country. Previous analyses have often focused on high-speed rail itself – seeking ways to account for its lack of support, funding, and adoption. However, high-speed rail is merely one part of a larger system, and in our view, the reasons why high-speed rail (and passenger rail transportation in general) may lack public support are reflective of larger systemic issues. These include the following:

1. A "silo approach" to transportation infrastructure planning and thinking, which applies to both government and business.
2. A transportation/infrastructure innovation deficit caused by lack of vision as well as a lack of funding to drive cross-silo transportation thinking, research, testing, and development.
3. Lack of economic discipline across modes, which is driven by public policy that subsidizes and insulates transportation users (particularly highway users) from experiencing and paying – at the point of purchase – the full cost of the road use service. This public policy subsidy gives the impression that vehicular travel is less costly than it really is, and it creates an especially high entry threshold for others who might wish to introduce alternatives.

We will address each of these three in more detail, but first, a disclaimer. The following ideas represent opinions and suggestions based on observation of – and reflection on – transportation systems, particularly those in the United States. Although detailed supporting documentation is not provided to support these comments and observations, these are shared with the hope that they will stimulate thinking and further discussion about new and unexplored transportation possibilities and will encourage the application of innovative imagination to this very complex and capital-intensive activity. Having said that, let us move on.

Government agencies that are tasked with transportation planning responsibilities have understandably focused largely on roads. Planning work in many such agencies centers largely on projecting and planning for growth in truck and automobile road usage. Other departments within the transportation planning arena handle issues related to rail, airline, and bus transportation, but the focus within each of these areas is often limited to addressing the needs of the infrastructure/mode assigned directly to it. Little opportunity or time, it seems, is afforded to considering cross modal or inter-connected modes of moving people and freight, or in weighing and analyzing different transportation infrastructure mixtures to determine what "blend" of infrastructures might deliver the best overall solution. This planning deficit may be a significant factor in limiting innovative developments and their application in the passenger- traffic movement business. Opportunities for cross or inter-modal discussion could provide more opportunity to ask critical questions and explore alternative infrastructure configurations or enhancements rather than taking the assigned infrastructure and thinking only of ways to expand or improve it.

For example, one question that we should be asking is this: could we reduce our need for roads – together with the related costs and ever-increasing congestion – if we looked at expanding other options in the infrastructure mix? In fact, could we improve the overall system if we considered the benefits of expanding other forms of infrastructure or of incorporating altogether new forms? Would a different arrangement and configuration of infrastructure alternatives reduce overall spending on transportation and/or foster enhancements in transportation quality? Do we even have a method to identify viable new alternatives, and to evaluate and test them? Do we know what enhancements might need to be made to other modes in order to make them attractive and viable enough to capture a desired percentage of passenger traffic?

Obviously, existing government-sponsored operating authorities that run bus and train systems are not expected to consider questions like these. If your task is to operate a bus system, you run busses. If your responsibility is a light rail system, you run light rail cars. If you are Amtrak, you run a train system. Most operating agencies are not expected to think of evaluating mode alternatives or to consider what mode, combination of modes, or blend of infrastructures should be used to move passengers most efficiently and conveniently. Managers of these systems have neither the time nor the mandate to do so.

The same impression is gleaned from a review of for-profit business entities that move passengers. They are usually focused on operating within their infrastructure silos too. If you are an airline, you fly planes. If you are Greyhound or Megabus, you move people on busses. If you are a ferry system, you operate ferryboats. Little or no thought seems to be given to alternative ways to deliver a point-of-origin to point-of-destination transportation solution – blending modes together and moving seamlessly from one infrastructure to another – thus eliminating the problems of multiple ticketing points, baggage handling, challenges to passenger movement, TSA security screening, etc. In fact, current "mode-transition points" (stations, stops, and platforms) have not changed much in the past century, and passengers are still expected to fend for themselves when leaving one mode/infrastructure and transitioning to another.

The question deserves to be asked whether there is a way to deliberately evaluate and plan for better multi-mode or cross-mode passenger movement. Our belief is that there could be – especially if we created the expectation and the opportunity for planning agencies and businesses to consider and explore these options. Because we have not yet developed ways to move passengers across modes more seamlessly does not mean it cannot be done. The lack of solutions may be a result of a lack of imagination and a dearth of good modelling and testing of new alternatives rather than the impossibility of the task.

One such cross-mode or multi-mode suggestion that the authors have developed has been more fully presented in another chapter in this book. That article explores the idea that air carriers must consider the potential benefits of collaborative arrangements with other mode operators. However, this collaboration should be developed in a way that allows them to manage the passenger and the passenger experience from end to end – regardless of the mode or modes used. This collaborative model would be more optimal and efficient rather than shifting or transferring passengers to competing non-air modes such as rail carriers or bus companies (See chapter “Exploring the Airline-High Speed Rail Collaboration Model: Efficient Services and Mutual Benefits.”).

Another seemingly unexplored innovation is designing a container that can navigate across modes. Shipping containers used for freight do move inter-modally of course, although admittedly the intermodal transfer methodology is a bit too “clumsy” to consider for passenger movement. Nevertheless, is there some reason why a more refined passenger container could not be developed that could allow the “container” or vehicle – with its contents – to move smoothly from roads to rail and back again? To explore this question further is there a reason why this container or vehicle could not be designed to be infrastructure agnostic; that is to say, capable of running on roads, rail and even new or underdeveloped infrastructures and able to move easily from one to the other? If so, would this not open up the passenger transportation industry to as-yet-unforeseen advances? By way of example, what if a monorail infrastructure were introduced that would allow some portion of the “container’s” infrastructure journey to traverse overhead through congested areas – or across rugged terrain – where road or rail capacity is not available or where alternative infrastructure costs are prohibitive? What if the system allowed for stations at second or third story levels in congested areas? It seems like these are the types of questions that could lead to possible alternative solutions to respond to our current and emerging congestion, capacity, and cost challenges.

While we have mentioned overhead monorail, allow us to further expand on this concept, since if offers one example of another option for a cross-mode or inter-modal infrastructure that could be selectively added to the transportation system. Assuming that the research and investigation indicated the feasibility and relative attractiveness of this alternative in a particular setting, the next step would be to build a demonstration monorail line that would address the need, and perhaps to design or modify a vehicle that could travel on more than one infrastructure. It is likely that this alternative might be particularly attractive in an area were other options are too costly or not feasible. The demonstration project would encourage public awareness and support for selective use of monorail and could demonstrate the attractiveness, functionality, benefits, and feasibility of such an alternative when compared to other options in the chosen setting. Project parameters could be set so that the project explores, considers, or addresses the following elements:

- Demonstrates both passenger and freight carrying capacity and automated mode-to-mode container movement.
- Enhances or complements existing public transportation infrastructure systems.

- Is designed as part of a larger system so that the value of the project grows as the system is expanded.
- Makes infrastructure system access available to both private and public entities. (The primary function of the system operators in this setting would be building and maintaining the infrastructure, just like a highway system, but the monorail system itself, as well as other infrastructure systems to which it attaches, would be available for both public and private use. The key is for the infrastructure operator to create and maintain the infrastructure and to set the expectations and rules for the use of the "road.")
- Is designed to accommodate automated container routing and switching that would move freight and passenger traffic to the correct destination. (We already automatically route electronic information packets efficiently and seamlessly across the internet infrastructure using various transmission modes such as copper wire, fiber-optic cable and wireless technology. The information packet is truly indifferent to what type of conduit or infrastructure it travels over. Is there some reason why a similar or analogous approach could not be applied and adapted to the movement of *physical* containers as well?)
- Seeks to benefit varied constituencies and socio-economic groups.
- Provides a means to generate sufficient revenues to continuously reinvest in the maintenance and improvement of the monorail infrastructure.
- Creates net positive societal benefits by allowing for a shift in the mix of infrastructure usage that reduces dependence on other less beneficial modes, given the assignment of reducing overall transportation costs and generating other societal benefits (such as reduced emissions, noise pollution, traffic congestion, commuting delays, and fossil fuel consumption).
- Appeals to the public sense of romance and fascination with transportation and creates a positive, memorable experience for passengers. This is the time to consider transportation aesthetics, safety and security, quality and convenience of service, luxury enhancements, choices and comfort.

A demonstration project would doubtless be expensive, but it may be less expensive than other alternatives – such as building more roads. If done well, it could lead society in general to see potential ways to leverage existing infrastructure systems and expand beyond them. And in the best of all possible worlds, if the project could be designed to produce a profit as well, it would represent a breakthrough in transit infrastructure economics and allow the means to plan for continuous expansion and improvement. As with all systems, we need to design our systems for the long term and to plan for the sustainability of the system, not just for getting it built. Finally, the development of a new infrastructure like this could provide business opportunities for companies that build transit conveyances as well as those that operate passenger and container transit businesses.

Our final observations focus on public policy considerations. It is our belief that public transportation policy should foster better economic discipline in the context of decisions about transportation purchases. To do so, public policy should endeavor to determine fully allocated costs for each transportation alternative, and those costs should be presented to the transportation consumer at the time of purchase.

Our assumptions are that transportation purchase decisions made on the basis of fully allocated costs would: (1) allow government to get out of the perpetual transfer-payment trap that it currently faces, (2) force society as a whole to fund the true cost of transportation purchases at the time they are made, and (3) encourage the introduction of new cost effective transportation options without having to overcome the current highway subsidization threshold. When consumers are presented with transportation decisions that reflect the full cost of their choices, they would be more likely to shift consumption patterns, and their purchase decisions would be made in the absence of the types of current subsidies that skew consumer decisions. Furthermore, full-cost decision-making permits cost transparency and the opportunity for providers of alternative options to offer those other options without the added hurdle of overcoming subsidized highway system advantages, for example.

Early thinking and revenue generation to fund the federal highway system funding followed a line of reasoning similar to this. The motor fuel tax initially functioned as a reasonable proxy for a user fee. However, much has been written lately about the diminishing effectiveness of the motor fuel tax to raise the necessary revenue to maintain, much less expand, the current highway system. With improvements in gasoline and diesel engine design and fuel efficiency, and the introduction of flex-fuel, natural gas, and battery operated vehicles, the fuel tax method of raising funding for roads is no longer as viable, or equitable, as it once was. In addition, as roadway construction and maintenance costs have risen, the motor fuel tax no longer generates enough revenue to build and maintain our roadways.

Finally, the politicized nature of the motor fuel tax interferes with rationalizing the means to pay for our highway system. Failure to increase the motor fuel tax as revenues diminish creates a system that does not have the revenue necessary to maintain itself, and it limits the ability to introduce other options. Ironically, whenever the price of motor fuel rises, some elected officials will call for a reduction in motor fuel tax to make fuel more affordable, a politically popular expedient but one that completely fails to answer the question how we are going to pay for the roadway system if we stop collecting the revenue we need. These trends and this type of mindset ignore financial reality, and the collective outcome is a public policy that hinders transportation innovation and that results in advocates for each transit mode clamoring in Washington for a bigger subsidy from the diminishing public pie.

Perhaps a food industry analogy will serve to illustrate the point we are trying to make. Our current to "road use consumption" policy provides road use services without any apparent direct cost. Except for toll roads, anyone can drive on most public streets and highways with no immediate cost. Let us suppose that grocery stores similarly would provide access to the food on their shelves without any direct cost. Anyone could walk into the store and take what they wanted. Of course, we know that the food distribution infrastructure would still need to be built and maintained. However, what if government decided to own and operate all grocery stores – just like government owns and operates all roads – and to pay for the expense of building and maintaining this food distribution infrastructure not by charging users for the food them consume, but by levying a tax on something they need in order to consume the food – say, a tax on dishes, eating utensils, and food waste disposal services? As long as the tax raised enough revenue to pay for the food distribution infrastructure, this system could work. However, similar to the experience with motor fuel taxes, it is fairly easy to play out various scenarios that could develop if this approach were taken. Consumers would be motivated to reduce their use of dishes and eating utensils, and they would likely switch from disposable products to permanent ones to lower their cost and to avoid paying the tax, and they would find ways to reduce or eliminate food waste. The result would be diminished tax revenue for the grocery distribution system. Additionally, there would be little incentive for politicians to risk political careers by raising taxes on dishes and eating utensils and food waste disposal (or gas taxes, to bring the point home), or even to suggest changing the policy so that consumers would pay directly for their groceries at the checkout lane when they leave the store. Politically, it would be much easier to take money from the public treasury and use those funds to "plug the gap."

Although admittedly this illustration is a "stretch," nonetheless it provides us with a rough analogy that helps us to understand why our current transportation policy gets us into trouble. When road use services appear to be free, determining how to raise the funds to pay for them is problematic. The public will be predisposed to use them predominantly because all other alternatives come with a price. Free roads have created a high hurdle for any other public transportation option to overcome. As illustrated with the example for funding the food distribution infrastructure, consumers have little incentive to consider other more costly options, and food producers would have little financial incentive to innovate new food products. In fact, some politicians would argue that producers' profits should be limited or that their products be subsidized directly from the public treasury. As a result food choices would tend to diminish and food quality would suffer. And importantly, there would likely be little – if any – food product innovation – an outcome very much like the current passenger transportation impasse.

The alternative to the full-cost absorption pricing is a continuation of this current subsidized transportation system. However, a review of history shows a pattern of increased government borrowing to fund public projects – borrowing that often lacks a clear revenue generation plan to repay the debt incurred. As the public appetite for subsidies of all sorts has grown, more interest groups are clamoring even harder for a shrinking portion of the public pie. The competition for public funds is already fierce, and this paper seeks to take one needed user service and remove it from this subsidy competition, where it is likely to suffer from chronic funding limitations, and place it on a self-sustaining trajectory.

The point is that society needs a market-responsive method to collect enough revenue to build and maintain not only the highway system, but also all present and yet-to-be-discovered alternatives. We must find a way to allocate that revenue fairly to existing and potentially new infrastructure providers. It would seem that presenting true costs to the consumer at the point of making a transportation decision would allow economic forces to shape transportation developments and encourage the introduction of innovations that might otherwise be delayed.

Therefore, it is our view that our country needs to develop a revenue generation method designed to collect the fully allocated cost of using any particular system at the time the purchase decision is made. A corollary to this alternative would involve the discontinuation of fuel taxes, since fuel tax is no longer an effective proxy for such user fees. A new methodology to collect revenue for roadway usage must be developed, and it should be rebased and updated at least annually with a view to generating enough revenue over time to maintain roadway system integrity without other federal and state tax subsidies.

What might an alternative system look like? Of course, user fees (tolls) are certainly one method to use where those can be efficiently collected. For most vehicular travel, however, we now have the ability to develop a GPS related tracking system which would allow us to bill for transportation in a way similar to cell phone billing, and that new billing system could replace the fuel tax as a revenue raising method. For example, each user could be billed on a per mile basis for road use based on factors such as the type of road used, the distance travelled, the time of day, the weight of the vehicle and load, and other variables. A system like this could bill the user more for traveling in congested corridors during peak travel times.

Ultimately, all costs for all roads could be determined – from the local street in front of a residence to the interstate highway. Using a GPS data collection system, which tracks mileage on all roads, as well as time of day, would provide a means to bill roadway users for their specific road usage patterns and to bill enough collectively to pay for the related costs of providing these users with access to those infrastructures. In fact, a system like this could also help reduce congestion by selective time of day pricing.

It is our belief that if the full cost of transportation decisions were more transparent and users were presented with realistic costs for their transportation purchase decisions, users would make economically-informed decisions about transportation choices and would consider other options more seriously, or at the very least, would encourage efforts to provide other less costly alternatives. Informed choice purchase decisions may also allow entrepreneurs to introduce new, cost competitive transportation alternatives that would not be viable in the context of the current subsidized highway system.

Regardless of the revenue generation method, society as a whole must raise enough funding to pay for each transportation option. The alternative to a "full cost" revenue generation policy is continued underfunding of the infrastructure, ongoing deterioration of what is already in place, ongoing battles to shift revenues from other sources to cover roadway costs, and higher hurdle levels for entrants offering new transportation options.

The underlying premise of this chapter is that the presence of transportation alternatives reduces the risk of economic shock if one option fails or faces a severe disruption. For example, the availability of other options reduces the economic impact of a fuel supply shortage, the collapse of a major bridge, weather disruptions, and the like. Users can adjust their travel plans by selecting other options that buffer them from the negative impacts of disruptions. A robust network of transportation options also provides business and leisure travelers with flexibility and serves to support economic growth and vitality.

The freight rail system operates efficiently, and the fully allocated costs of operation are paid for by system users. Government does not, to use another analogy, build and maintain the freight rail infrastructure and then attempt to raise the revenue to pay for this system through a tax on freight rail diesel fuel sales – a tax that would only raise 50% or 60% of the total cost of the system. No, rather the full cost of the freight rail infrastructure is borne by those who use the system through various charges for services provided to freight rail system users. Similarly, for many of the reasons already outlined in this paper, would it not make equally good sense for the cost of road use to be borne by those who use the roads?

One issue that often surfaces is the challenge of providing reasonably priced transportation options for those who are economically disadvantaged. The argument could be made that full cost allocation pricing of public transportation would make public transportation less affordable for this group, since they are more likely to use public transportation options. However, if a fully allocated cost approach lowered the entry threshold for other public transportation entrants or shifted enough revenue into existing options to allow for improved service, the economically disadvantaged would certainly benefit. If needed, a more "surgical" strategy could be designed to assist individuals with a transportation benefit program, which could be tailored to specific, individual needs. Similar programs have been developed to provide food assistance without subsidizing for-profit grocery stores. However, if public transportation systems attracted more riders because other full-cost options were more expensive, the added revenues would reduce the need for consideration of subsidies of any sort, and the ultimate goal, as this paper suggests, is to provide options that are self-sustaining.

In summary, it is easy to get lost in the details of transportation planning, and obviously, this article has offered detailed ideas intended to illustrate possible solutions to the three reasons suggested for the lack of good rail transportation alternatives. We should remember, however, that the broader purpose is to highlight the negative consequences of "silo" thinking and planning. This silo approach is reflected in both the public and the private sectors and may be a primary factor hindering true transportation innovation breakthroughs.

In addition to challenges created by silo approaches to transportation planning and development, public transportation policy also needs to be significantly revised. New policies need to be developed to allow transportation changes to evolve and respond quickly to changing constraints of all kinds. Without a policy that allows the transportation system to respond and adjust gradually to changing constraints, we are likely to trap ourselves into an endless spiral of ever-increasing subsidies for one form to the detriment of other promising alternatives.

REFERENCES

Dumaine, B. (2015, June 4). Why America is betting big on bullet trains. *Fortune.com*. Retrieved from http://fortune.com/2015/06/04/americas-bet-on-bullet-trains/

Federal Railroad Administration. (2014, September). *Southwest multi-state rail planning study: Summary report,* 1-11. CrossRailChicago: A vision for a modern Metra. Retrieved from http://CrossRailChicago.org

Federal Railroad Administration. (2015, August). *Dallas to Houston High-Speed Rail Project Corridor Alternatives Analysis Technical Report.* Retrieved from https://www.fra.dot.gov/eLib/details/L16978

Kanter, M. (2015). *Move: Putting America's infrastructure back in the lead.* London, UK: W.W. Norton & Company. Retrieved from http://www.amazon.co.uk/Move-Putting-Americas-Infrastructure-Back/dp/0393246809

Kelly, A. (2016). Maglev Train: USA gets $28M study for 300 mph train travel that gets from D.C. to NYC in less than an hour. *IDigitalTimes.com*. Retrieved from http://www.idigitaltimes.com/maglev-train-usa-gets-28m-study-300-mph-train-travel-gets-dc-nyc-less-hour-490793

Kelly, B., & Nichols, M. (2014, January 29). Taking California's bullet train to a greener future. *LATimes.com*. Retrieved from http://www.latimes.com/opinion/op-ed/la-oe-kelly-california-high-speed-rail-20140129-story.html

Selladurai, R. (2013). Indiana high-speed rail: 'LINK' to the future. *The South Shore Journal*, *5*, 171–182.

U.S. Census Bureau. (2011, March). Population Distribution and Change: 2000-2010.

Section 3

Chapter 11
Progress through Partnerships:
The History of the Northeast Indiana Passenger Rail Association

Dorothy Hagan
Independent Researcher, USA

ABSTRACT

This chapter provides a brief history of the development of the Northeast Indiana Passenger Rail Association (NIPRA) and includes: 1) An overview of two studies utilized by NIPRA to develop their preferred rail corridor through Fort Wayne; 2) the partnerships that were developed at the local level across the corridor; 3) the study completed to provide a business case for the corridor; 4) discussions on coordination and fundraising efforts by NIPRA and the City of Fort Wayne staff, including a Federal Railroad Administration (FRA) grant application; and 5) the next steps for NIPRA and corridor partners.

INTRODUCTION

The Northeast Indiana Passenger Rail Association and the City of Fort Wayne are leading the effort to bring passenger rail back through northeast Indiana while connecting Chicago, Illinois to Columbus, Ohio. The partnerships of this effort has progressed far enough that the Indiana Department of Transportation (INDOT)

DOI: 10.4018/978-1-5225-0102-2.ch011

sponsored the submission of a grant to the FRA for funding the required Tier 1 Environmental Impact Statement (EIS). The chapter provides a history of NIPRA and its partnerships.

BACKGROUND

This chapter has been informed, reviewed, and edited by members of the NIPRA Board and staff of the City of Fort Wayne Planning Department. As this is only a historical essay, there are no literature reviews or "positions" taken for this chapter.

The Northeast Indiana Passenger Rail Association is an organization, which has been dedicated to bringing rejuvenated passenger rail service to Fort Wayne and Northeast Indiana. The organization believes that our transportation system should include rail, highway, and air modalities. According to Paddock and Hayhurst (2009), passenger rail service would provide several benefits to the residents of Northeast Indiana and surrounding areas including a stronger economy, increasing variety of jobs, enhanced security, and a cleaner environment.

The mission of NIPRA is to collaborate and work closely with the local, state, and federal elected and appointed officials, as well as business leaders in this region, to promote the bringing back of passenger rail service to Fort Wayne. This service would be in addition to the train service that currently connects through Waterloo. Also, NIPRA hopes to present more actively the refurbished Baker Street Train Station, the growing citizen support, and the positive community enthusiasm that has been generated for passenger train service in and around our city. NIPRA contends that the state of Indiana intentionally and actively collaborate with the surrounding nine Midwestern states to bring a sophisticated and efficient passenger rail service network to the Midwest, which would provide major economic and other related benefits for Fort Wayne and Northeastern Indiana too. Experts have estimated that about 4500+ full time permanent job would be created in Indiana alone with the completion of the high-speed rail network (Paddock & Hayhurst, 2009).

One of the major purposes of NIPRA is to focus on positively influencing the local government and business leaders in the region that supporting the reinstitution of passenger rail service in Fort Wayne will benefit our city and state in many ways. We point to some obvious facts with our current transportation system. Our highways are becoming highly congested with increasing traffic, and our nation's airports are becoming extensively overloaded with travellers in an increasingly mobile society that an optimal alternative like high-speed rail becomes highly critical, attractive, and necessary. Passenger trains can provide efficient and timely travel, particularly with destinations from 100 to 300 miles in length. Although gasoline prices are currently relatively low, prices could easily climb to the $4+ per gallon mark again, and fossil fuels are the major cause of air pollution. Rail transportation uses approximately one tenth of the fuel per passenger mile, when comparing a train with 400 passengers with an automobile carrying one passenger. Hence, when we can move more travellers via passenger rail transport modes, this will lead to reducing our dependence on foreign oil and bring us closer to the eventual goal of energy independence. Further, when airplanes are grounded by inclement weather conditions or a national emergency, train travel would provide a ready and efficient alternative means of moving stranded passengers throughout the country.

THE EARY YEARS

Indiana has been active in Midwest passenger rail development since signing as one of the original members of the Midwest Regional Rail Initiative (MWRRI) in 1995. Citizens of Fort Wayne became involved in the initiative to bring passenger rail to back to their city in early 2000. The DeKalb County, Waterloo, and Auburn representatives serving on the Northeastern Indiana Regional Coordinating Council (NIRRC) and the local Metropolitan Planning Organizations[1] (MPO) expressed that their communities did not like the rail stop in Waterloo and requested the route return to Fort Wayne. After hearing this, local NIRRC representatives, including Geoff Paddock (current Fort Wayne City Councilman) decided to step up efforts for bringing passenger rail back to Fort Wayne. Paddock was originally Governor Evan Bayh's appointee to the NIRRC Board, and the late Governor Frank O'Bannon reappointed him. Paddock and other NIRRC representatives approached Governor O'Bannon and the Indiana Department of Transportation (INDOT) for assistance with this initiative. In 2002, INDOT, along with Ohio Rail Development Commission and Amtrak, co-sponsored the routing study, the Northern Indiana/Northwestern Ohio Routing Study, which resulted in Fort Wayne being selected as an intermediary stop on a proposed high-speed rail line between Chicago and Cleveland. This more southern route eastward was chosen as part of the final Midwest Regional Rail System plan adopted in 2004.

Unfortunately, for Fort Wayne, political push back arose from South Bend Statehouse representatives, who also wanted this high-speed passenger rail stop. These political pressures resulted in no line being officially promoted by the State of Indiana, and with the economy failing in 2007, discussions of any passenger rail expansions weakened, as government at all levels struggled to just meet basic needs. Then in 2009, the American Recovery and Reinvestment Act (ARRA; P.L. 111-5) became the new federal funding source that would reinvigorate the development of high speed passenger rail (HSR) transportation throughout the United States.

The Northeast Indiana Passenger Rail Association (NIPRA) originated as an ad hoc advocacy alliance sponsored by Invent Tomorrow, an organization formed in 2000 to tackle Fort Wayne's economic development issues and improve the community's image. In 2009, recognizing the substantial economic benefit of re-establishing passenger rail service through northern Indiana, the City of Fort Wayne City Council passed a resolution supporting the return of service. Also in 2009, NIPRA incorporated and established a 15 member board consisting of representatives of private industries, municipalities and elected officials. At this time, Mayor Thomas Henry directed the City to provide funding and staff resources to support the work of NIPRA.

Fort Wayne rail supporters knew they needed to make a concerted effort to bring passenger rail back to their city, and in 2009, INDOT became a signatory to the Memorandum of Understanding among eight (8) states and the city of Chicago, agreeing to seek and coordinate funding from the American Reinvestment and Recovery Act (ARRA) and beyond. INDOT's application for \$2.8 billion of ARRA included funding for the Tier 1and Tier 2 EIS for the Chicago-Fort Wayne-Toledo and on to Cleveland route of the MWRRI, construction dollars for the Chicago-Cleveland corridor, and money to upgrade the Chicago-Indianapolis-Cincinnati line to 110mph. This sizeable proposal was not awarded any money (Figure 1).

Figure 1. NIPRA Map
Indiana Department of Transportation (INDOT), 2009

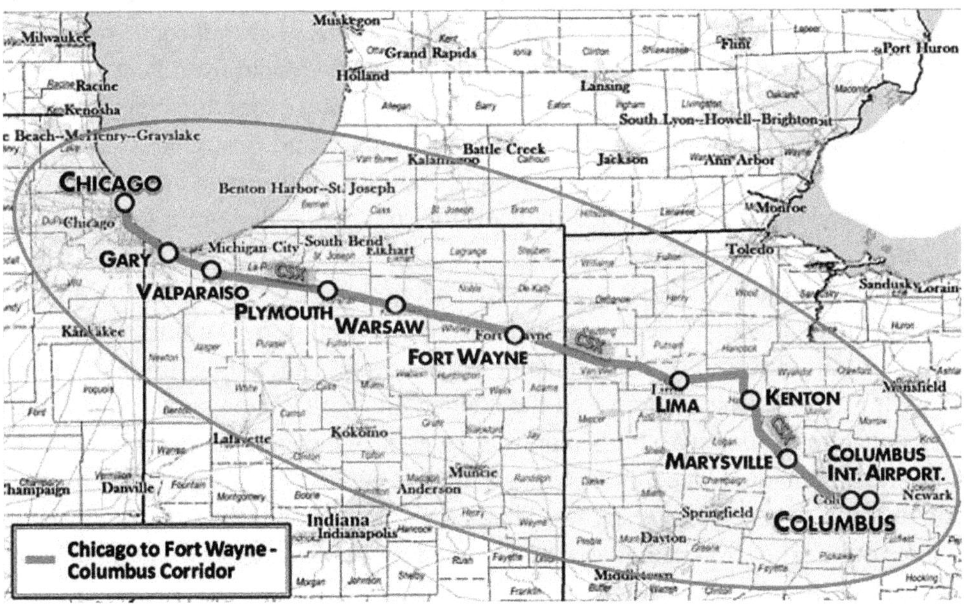

The Chicago-Fort Wayne-Columbus Corridor is a unique rail initiative because it has developed from the bottom up rather than the more traditional top down approach. Cities and local communities along the route have led the effort in promoting and fundraising for the business analysis and EIS studies. Rail development has traditionally been spearheaded at the state level; with the federal government approached later for assistance. Local governments may have money in station development but have rarely been involved at the onset of the effort.

This bottom up approach has both benefits and risks. The main benefit is the broad base of support for the corridor developed even before seeking state or federal money. With already hundreds of supportive fans, including many businesses and educational institutions, higher-level government officials take notice and are compelled to give their support. The risk of this approach could happen during the federal government required EIS process as the current method analyzes the environmental impact of new routes between two designated endpoints. This method could impact all of the projected "passenger rail stop" communities along the Chicago-Fort Wayne-Columbus Corridor who have already funded or committed to funding this particular route. The traditional analysis may need to be altered to allow for the stops within these communities. This Chicago-Fort Wayne-Columbus Corridor is not to this point in the process but will be very soon.

GEOFIGUREAL ROOTS

The Chicago-Fort Wayne-Columbus Corridor actually has its geofigure roots in two prior regional rail planning initiatives. The MWRRI lists a route from Chicago through Fort Wayne and the Ohio Hub Plan lists a route from Fort Wayne to Columbus. Individually these plans identified the strong economic benefit of connecting rail service from Fort Wayne to both Chicago and Columbus. Relying on these studies for support, NIPRA used this opportunity to expand their role from advocate to leader and expanded their efforts to develop the entire Chicago-Fort Wayne-Columbus Corridor (Corridor).

The first of the two studies, the MWRRI network, was completed in 2004. Planning had begun nearly a decade earlier with collaboration by Amtrak, the Federal Railroad Administration (FRA), and nine Midwest states, including Illinois, Indiana, and Ohio. This initiative proposed the Chicago to Fort Wayne corridor, although the connecting Ohio city on this plan was Toledo. A second regional initiative, the Ohio Hub Plan, developed as an adjunct to the MWRRI and focused on Ohio rail connections. One corridor designated in the Ohio Hub Plan is the Columbus to Fort Wayne route. The Ohio Hub Plan was approved for Federal ARRA funding in 2009, but Ohio Governor Kasich did not support passenger rail development at the time and this eventually led the FRA to recoup the funds from Ohio.

The approximately 315-mile corridor from Chicago, Illinois, through Fort Wayne, Indiana and on to Columbus, Ohio is essentially a combination of the Chicago to Fort Wayne section of the Chicago Hub Network presented in the MWRRI, and the Fort Wayne to Columbus section of the Ohio Hub Plan. The Corridor will link to the Chicago Hub Network, where substantial investment has already been made in developing the MWRRI to the neighboring states.

THE PARTNERSHIP

The forging of the partnership between the City of Fort Wayne and NIPRA would solidify efforts for the development of a Chicago-Fort Wayne-Columbus route. At a "Rally for Rail" held at refurbished Baker Street Station, hundreds of people including numerous government officials joined together to inform Derrick James, a senior government affairs officer with Amtrak, how much they wanted passenger rail back in Fort Wayne. Then, at the urging of then Indiana Governor Mitch Daniels, and satisfied with the progress of their initiative, NIPRA, Fort Wayne and Columbus joined to test the initial route feasibility.

The group contracted with Transportation Economics and Management, Inc. (TEMS), to complete a business impact study. The study was paid for with donations from a variety of sources including local governments, businesses, and foundations. The TEMS study process began in 2011 and the report was completed in December 2012.

Also during 2011, Mayor Tom Henry began a community initiative, Legacy Fort Wayne, which allowed Fort Wayne residents to have a voice in identifying potential "transformational" projects for improving the community. The return of passenger rail rated as a high priority in the community and set the stage for the City of Fort Wayne to pledge the first major contribution, $200,000, to the funding of the Tier 1 EIS. This first pledge played a crucial role in convincing other funding sources throughout Indiana and Ohio that Fort Wayne and NIPRA are seriously committed to return passenger rail to the city.

THE BUSINESS CASE

The TEMS study originally identified projected stops along the route to include Chicago in Illinois; Gary, Valparaiso, Plymouth, Warsaw, and Fort Wayne, in Indiana; Lima, Kenton, Marysville, Hilliard and downtown Columbus, in Ohio. However, during the final stage of the study, the City of Columbus pushed for an adjustment of the route. Columbus proposed the elimination of the stop at Hilliard, Ohio and introduced a stop at Port Columbus International Airport. The TEMS technical analysis for the section from downtown Columbus to the Port Columbus Airport was never completed due to lack of funding. An Executive Summary completed in February of 2013, did eliminate the Hilliard stop and added Port Columbus Airport, as requested by the City of Columbus. In spring of 2014, Columbus Mayor Coleman contracted for a study to complete, among other tasks, a feasibility analysis of the downtown Columbus to the Port Columbus International Airport portion of the corridor.

After the completion of the TEMS Study, NIPRA and Fort Wayne city staff shifted to a more active role and began developing strong partnerships with representatives of the Indiana and Ohio communities along the corridor. To facilitate continual dialog in Indiana, NIPRA has coordinated quarterly meetings with the Indiana mayors, community representatives, congressional and senatorial staff members representing the corridor, MPO staff and other key stakeholders. Additionally, several meetings involving the entire corridor have occurred and will continue to occur as the initiative progresses. Individual mayors along the corridor have generally hosted these meetings with NIPRA members and City of Fort Wayne staff assisting with the invitations, agendas and personal phone calls to ensure maximum attendance. Mid-Ohio Regional Planning Commission (MORPC), the MPO for Columbus Ohio, has been instrumental in coordinating efforts throughout Ohio.

COORDINATION AND FUND RAISING

Meetings throughout the corridor have been instrumental in keeping the process moving and holding all the partners accountable. These meetings also ensure that the partners maintain a consistent and productive message about the project and that the leaders of the impacted communities are up-to-date on all accomplishments. These meetings have been a platform for discussing topics including the establishment of the Memorandum of Agreement (MOA), fundraising strategies, and tactics for creating a larger network of supporters. The next step in the corridor development process for which NIPRA has been strategizing, is to further organize the partnerships, define roles and plan for raising funds for the legally required Tier 1 Environmental Impact Statement.

Although all partners agreed on the principal need for an MOA, the actual production and execution of the document would test NIPRA, Fort Wayne staff and MOPRC's ability to coordinate, collaborate and implement joint processes among the corridor members. With an initial projected signing deadline of spring 2014, the effort to accommodate all necessary changes by all partnering cities substantially prolonged its completion. Draft changes included adding small sections of detail about each city with a rail stop, adding language about coordinating fundraising using a mutually agreed-upon strategy, and adding a section about future signatories to the MOA. All partnering cities eventually signed the MOA by September of 2014. Additional work is being done to include signatures of county executives, INDOT and Ohio DOT staff, and the five MPO's along the corridor on the MOA.

When fundraising for the Tier 1 EIS began in earnest, all involved NIPRA and City staff members contacted educational institutions, not-for-profits and private entities to gain support for the corridor and raise funds for the Tier 1 EIS. At a meeting late in 2013, NIPRA and City staff members offered up their services to identify, contact, and meet with possible funding contributors along the Indiana portion of the corridor. NIPRA members traveled hundreds of miles throughout the Corridor making presentations and successfully obtaining over $450,000 in pledges and contributions.

This procurement of funds placed the progress of the corridor in an optimal position to apply for an FRA grant, which was an unexpected opportunity that arose in the fall of 2014. If received, FRA funding will speed up the process of completing the Tier 1 EIS, as the program allows an 80/20 arrangement with the federal government funding the 80%. The remaining 20% was pledged equally from the Indiana and Ohio partners. NIPRA and supporting staff members are hopeful that early and consistent contact with the regional FRA program manager, along with the presentation given by MORPC to FRA Administrator Joseph Szabo, will improve the Corridor's chances of being awarded grant funds.

FRA GRANT OPPORTUNITY

Recognizing that the role of the INDOT may be minimal in the beginning stages of generating and organizing the corridor's support, NIPRA still insures that they remain in contact with INDOT's rail staff throughout the progress, sending email updates, inviting them to meetings and conference calls and personal phone calls when necessary. This partnership was crucial when the FRA grant opportunity arose, as the State was the only entity eligible and willing among the Corridor partners to apply for the grant. After agreeing to be the signatory, INDOT requested NIPRA and City of Fort Wayne staff members to draft the application and gather all supporting documents. As of this writing, it is not known if the grant application will be successful but regardless of the grant outcome, the corridor partners will continue in their efforts to build the Chicago-Fort Wayne-Columbus Corridor.

With the knowledge that the application had to demonstrate coordination and organization throughout the entire corridor, NIPRA and City staff members used this opportunity to encourage progress from the partners in Ohio. After the fallout of the failed Ohio Hub Study and feeling the constraints of political pressure from constituents and the Ohio Governor's office, the City of Columbus Mayor had been reluctant to weigh in on the project. Within the final hours before grant submission, and while remaining publicly quiet about his support, Columbus Mayor Michael Coleman signed the MOA and provided documentation committing the City of Columbus to raising their share of funds for Ohio's local match if the federal grant should be rewarded.

THE FUTURE

Future work needed includes discussions on project oversight of the Tier 1 EIS, decisions on a management entity for the corridor itself, project oversight and financing decisions, and even the naming of the Corridor.

MORE PARTNERSHIPS

Since 2009, NIPRA has also been supporting the further development of the Detroit-Chicago rail corridor as they work toward higher speeds and additional train sets. This project is currently undergoing the Tier 1 EIS, under the management of the Michigan Department of Transportation (MDOT). A portion of the corridor, from Gary into Chicago, is proposed to serve as part of the Chicago-Fort Wayne-Columbus route. In their EIS analysis, MDOT indicated that they would include the rail traffic of the Fort Wayne route from Gary, Indiana into downtown Chicago.

NIPRA and the City of Fort Wayne have also been meeting with rail advocates from Toledo as they seek to develop the Chicago-Fort Wayne-Toledo connection. This initiative is not yet as strong as the work with Columbus but it is being pursued as another great economic boost for northern Indiana communities. NIPRA believes there is strong support in Fort Wayne and Northeastern Indiana for the return of enhanced passenger rail service through the Baker Street Station. Recent data indicates that reviving a passenger rail line from Fort Wayne to Chicago and Cleveland or Columbus, Ohio would generate enough passengers to make it financially feasible. In addition to several major benefits, building an enhanced passenger rail system would create thousands of permanent jobs in Indiana. Steel needed to produce or replace rail lines could be manufactured at SDI (Steel Dynamics Inc.), creating additional economic activity in Fort Wayne and Northeastern Indiana. Further, the economic recovery legislation signed by President Barack Obama on February 17 appropriates $9.3 billion to further this cause. Eight billion dollars has been appropriated for capital assistance for high-speed rail corridors and intercity passenger rail service. This means there are now dollars available for states to advance the goal of new intercity rail systems across the country. Governor Mitch Daniels and state legislators must work to see Indiana is included in this system. It is imperative the state develop a rail plan to designate where funds are to be spent to improve and add Amtrak service, and Fort Wayne must be included in this plan.

CONCLUSION

We have presented the historical highlights of passenger rail related plans, efforts, and activities in Fort Wayne and Northeastern Indiana travel corridors. The history of NIPRA is not yet over, and it is continuing along as corridor partners are working diligently and collaborating their efforts to return passenger rail to northern Indiana and Ohio.

REFERENCES

Indiana Department of Transportation (INDOT). (2015). Retrieved from http://www.in.gov/indot/

Paddock, G., & Hayhurst, T. (2009, March 18). Northeast Indiana Passenger Rail Association (NIPRA) is formed guest. Retrieved from http://www.thewaynedale-news.com

ENDNOTE

1 Metropolitan Planning Organizations were created by Congress to insure that region-wide plans were developed through intergovernmental cooperation and rational analysis.

Chapter 12
Are We There Yet? Perspectives of a Pioneering Visionary

Dennis Hodges
Indiana Passenger Rail Alliance, USA

ABSTRACT

This chapter addresses a vision of what was, is, and can be. Japan is already 65 years into its passenger rail improvements and modifications. France is just over 35 years and China is eight (8) years into its passenger rail improvements and modifications. Why cannot the United States see the economic and environmental reasons for doing the same in this country? This chapter discusses that and more. The author has attempted to parcel out the whys, what's and where's as to how the United States became so parochial when it comes to passenger rail transportation; why it is so reticent towards moving forward, and what can be done to change the current political and social attitude and climate.

INTRODUCTION

A child traveling with adults may at some point, blurt out, "Are we there yet?" Depending on the mood of the adults, that question can be either humorous or annoying. It is the same with high-speed rail. "Are we there yet?" This question seems to be a qualified response to the many years of advocacy, planning, proposing, and studying from the perspective of a high-speed rail pioneer.

DOI: 10.4018/978-1-5225-0102-2.ch012

When I first rode the trains of Europe in 1978, high-speed trains were in their infancy while traveling about 125 MPH. In fact, I rode one such inaugural train in 1980, from Paris to Marseille. Today, the trains are reaching over 200 miles per hour, and even higher speed trains are likely to be introduced very soon. Now, a traveler can whisk from London to Paris to Brussels to points in Germany and many more European locations via high-speed trains. Almost all of Europe is saturated with rail systems that can get the traveler to where he or she wants or needs to be almost as fast as one could fly. In some cases, just as fast!

Europe is already there and clearly in the 21st century. A few years ago, I had the opportunity to visit locations where these dreams are being formed and implemented: Alstom of La Rochelle and Strasbourg France; Talgo of Madrid Spain and Siemens of Krefeld and Munich Germany and of Vienna and Graz Austria. I was invited by the SwissRail Association to tour the trains of Switzerland that took me high in the Alps and through some very breath-taking scenery. The Swiss, as do all Western European management groups, pride themselves on being precisely on time. Further, most high-speed trains meet/match other forms of transportation such as buses, commuter rail, and taxis with precision.

At the Siemens Test Center at Wegberg in Northern Germany, Director Robert-Andre Grootings gave me a Certificate of Completion for having "driven" a high-speed train. I learnt of the capabilities of the Siemens' ICE train and about the history of four generations of trains.

In Austria, I became aware of the Siemens' Railjet as the train that opened up passenger rail service from Europe into Hungary and Turkey. I really like this train for its style and speed. It only reaches speeds up to 150 MPH, but is perfect for what we should consider for Indiana and the Midwest.

In this regard, Armin Kick, Director of High Speed Rail at Siemens has stated that the Sacramento-based company is producing a rail product similar to the RailJet, and that the American version is PRIIA compliant (SIEMENS, 2010). Other companies, Alstom, Bombardier and Talgo are also rolling out similar versions of the concept, but I have not yet had the pleasure of evaluating them. However, the trains that Alstom and Talgo make for European transportation needs equally impressed me.

HIGH SPEED RAIL DEVELOPMENTS IN INDIANA

Having founded and organized the Indiana High Speed Rail Association in 1992, I brought attention to the possibility that high-speed rail needs to be considered for 21st century transportation in Indiana and the Midwest, and that it should be an integral part of the area's transportation vision. If Europe can embrace, envision, and implement this system, why cannot we do the same in the United States?

Unfortunately, the new movement has encountered serious opposition from the highway lobbyists, which nearly ended our vision. Having regrouped in recent years, the new Indiana Passenger Rail Alliance (IPRA) is determined to prove that the newer 21st century trains will not be a serious competition to highways or even airports. Rather, the trains are to be a part of a transportation system that includes and compliments all forms. Highways, especially around major urban areas are becoming more congested and airlines are looking to reduce, if not end, their short haul (commuter) flights. For example commuting from Indianapolis to Chicago, via highway and air, could take up to four hours. A high-speed train, traveling at 125 MPH or more, can get the traveler there in under three hours.

During their time on the train, the travelers can do office work, prepare for an appointment, read a book or magazine, have refreshment or simply relax in comfortable seats – all while traveling safely and enjoying the speed that gets them to where they want to be on time.

So, why cannot the appropriate commerce and government leaders see the benefits? There are several reasons. First, travelers have become very comfortable in their cars, even though they could come up to a traffic jam or backup to delay their drive time considerably. Also, more comfortable traveling by air may be another reason even though the necessary security measures are costing them much time and money.

Secondly, we have leaders that can only think in terms of concrete. If concrete is not being poured on to the soil, the rail project, in their way of thinking, is not worth considering. So then, this raises the issue of political will.

Passenger Rail does not yet have the voices that other modes of transportation have. However, that is changing. Various county commissions and municipalities are coming "on board" to support a newer passenger rail system. In addition, as more proof and studies point out the overall economic benefits, more and more leaders of commerceare chiming in support of such a system.

The Indiana Passenger Rail Alliance is going several steps forward by engaging the Transportation Economics and Management Systems, Inc. (TEMS) to do an Economic Impact Study and Business Plan. This follows similar work done by our sister organization, the Northeast Indiana Passenger Rail Association. Combined, these studies should provide added support for all the evidence that has been presented and will confirm the need for modern passenger rail systems to come on line.

There is also the issue of transportation competitiveness and attractiveness in the international market. By getting especially business travelers to their destination quicker and on time, the whole area of world trade in the United States can become more competitive and sophisticated.

Why should a tired or beleaguered traveler be delayed for a meeting or appointment? Why must we endure traffic issues or airport concerns, when a more modern rail system can deliver a visitor that has a more attractive presence? Speed is just about everything in this society, but so is humanity. Travel by rail delivers both.

I know we have many legislators and other elected officials that are looking at the bottom line – and they should. How much will it cost? To be honest the building of a 21st century passenger rail system is not cheap. It will cost states and regions of this country several billions of dollars. Yet, the United States has a real need to upgrade and modernize all infrastructure projects.

It might be surprising to note that this country does not have a transportation policy, nor do most of its individual states. Projects are built on an "as needed" basis or when an emergency arises. That is another reason why the newer trains have trouble getting attention. If our states and nation did have such a policy, it would put the trains into a broader context, and and allow for higher speed trains to fit in such a policy.

When I first organized what was then the Indiana High Speed Rail Association, I had the European models in mind and that because they, along with Japan, were the only ones providing the much-needed transportation service.

Today, the Indiana Passenger Rail Alliance (IPRA) is continuing to work on commissioning an objective business case study for higher speed rail in this state. The study will explore alternatives for funding and operating the system and will analyze operating revenue and expense models. The intent is to help decision makers look at the business opportunities and risks that a major investment in higher speed rail would create. People and cargo can be moved from place to place in several ways. A goal of IPRA is to create a planning mindset that evaluates the long- term cost-benefit ratios of the options at our disposal in order to maximize overall benefits in the long term. Building roads and operating cars and trucks on those roads is expensive, and will become even more expensive in the future. Studies of travel patterns where competitive high-speed rail options are offered show that an effective, well-developed passenger rail network may significantly reduce the demand for more roads, cars, and trucks (Indiana High Speed Rail Association, 2015).

However, things have changed and most of those states considering 21st century passenger rail service have now moved to a slower 110 to 130 MPH. That is OK! This generation of trains is designed for comfort, safety, and on-time performance. Almost all of the products presented for this purpose are visually appealing and communities would be proud to have them come to their particular city or town.

However, the trains are not the only attraction. Sleek, modern, and business-friendly passenger stations will also be inviting and highly appealing with the all the amenities and conveniences most travelers require. These stations would then foster commercial and retail developments, which will add to the economy of those communities.

This was largely true in the "hey-day" of passenger rail. In fact, there was a romance about trains that lasted for decades and that was never afforded other modes of travel. People really enjoyed the comfort and convenience of those trains. Then came the intercity highways and air travel that was marching into prime time. Subsequently, companies like the Penn Central, Santa Fe and New York Central abandoned their services to where the United States had to pick up the slack.

In 1970, the US Congress formed the National Passenger Rail Corporation and the new Amtrak service began operating in May 1971. Slowly but surely, passenger rail corridors began to disappear leaving communities like Fort Wayne, Indiana without a viable connection to major market cities. Today, Amtrak is a skeleton of what it was in the 1970's.

In 1986, there was a glimmer of hope that could make rail travel more appealing. Trains on the Northeast Corridor from Washington to New York were allowed to travel 125 MPH. That was considerably faster than the 79 MPH limit established by Congress in 1947. Then in December 2000, true high-speed rail came to America when Amtrak launched the Acela, a passenger train along the Northeast Corridor that could travel up to 150 MPH.

Today, plans are being rolled out in other regions of this country to accommodate sleek and modern passenger trains traveling at speeds of 110 to 150 MPH. In the Midwest, planners are looking at passenger trains traveling 110 MPH between Chicago and Detroit and Chicago and St. Louis. In Indiana, the Alliance wants the impending data to reflect the benefits of trains going 125 MPH or more.

In 1996, nine Midwestern States: Illinois, Indiana, Iowa, Michigan, Minnesota, Missouri, Nebraska, Ohio and Wisconsin agreed that 110 MPH should be the norm for its proposed passenger rail corridors. Since then, some of these same states are seeing even greater economic advantages of trains running at 125 MPH or more. The greatest cost of this form of transportation is in building the infrastructure. Most states wanting higher speed trains are looking at $6 to $10 million per mile, with the United States government paying about 80 percent of the tab. However, that is considerably less than their cost of building a new interstate highway (Indiana High Speed Rail Association, 2015).

BENEFITS OF HIGH SPEED RAIL

Now rail travel is more than money, mortar, and steel. It is really about PEOPLE -- the ones who travel and the ones who are looking for options and convenience. The proposed Midwest Regional Rail System is a transportation project that will significantly improve the level and quality of transportation services in Indiana and the Midwest. The system will also contribute to the region's economic growth and tax base (Midwest High Speed Rail Association, 2015).

How is that possible? All forms of transportation add dollars to the local economy by way of new retail and commercial developments in the periphery of their destination points, and passenger rail is no exception. Industries are built around transportation destinations and area tourism could see a boom in people coming to local events and historical sites.

When I travel in Europe, I take advantage of what the passenger rail system can provide. When I go to Paris, Milan, Madrid, Brussels, London, Munich and Berlin, the tourism part of my travel is satisfied immediately. I am right in the heart of history and unique architecture, as I am for such events as Oktoberfest in Germany or Bastille Day in France.

Similarly, the proposed trains for this region of the United States can do the same thing – take travelers to places both known and unknown, but there is a caveat. The emphasis of the Midwest Regional Rail System should be on the noun "system." The Webster Dictionary defines the noun as "a group of interacting, interrelated, or interdependent elements forming a complex whole."

Included in the implementation of these trains should be an integrated connection of buses or regional trains or available taxis and rental cars. A concierge needs to be at the station to guide travelers to their destination. Moreover, at the station, there needs to be plenty of amenities to accommodate both the business and leisure traveler. That ushers in another benefit.

The business traveler could enjoy the convenience of being able to prepare for an appointment while traveling the trains, make phone calls, and even conduct business on the train. For both the business and leisure traveler, they would enjoy a greater ease of movement or a refreshment in the lounge or dining car. The use of electronic devices would not be a problem for those traveling this mode. Why would anyone not want to travel by rail? It just makes common and economic sense.

How safe are these trains? It is a common question that enjoys a great positive answer. In its near 70-year history, there have only been a couple instances where the modern trains had a problem. However, engineering has remedied those concerns and the safety record is virtually flawless, as is their record of on-time schedules.

In a study reported by Richard Florida (2009) in the *Atlantic Monthly*, delays due to auto congestion cost the American economy $81 billion. Included in this figure are the missed appointments that were previously scheduled. High-speed trains will reduce the stress of harried travelers that becomes typical with highway gridlock and airport delays. Add it all up, and passenger rail can relieve travelers from all the issues that come from other modes.

I once had a vision where something like a "Silicon Valley" could spring up around the rail corridors. I still believe that to be true. Once the corridors are established, industries of all types will spring up along them. Local and regional planners can market the respective properties to accommodate specific industries. In fact, I would suggest that modern, 21st century passenger rail would clearly accelerate commercial, industrial, and retail growth.

This, of course, means many jobs for just about every skill level. The rail stations could become anchors for such developments, and a magnet for urban renewal. Tourism sites and activities, especially those underserved by public transportation, will see a measurable increase in visitors. In Indiana, several companies could build or contribute to the building of train locomotives and passenger cars. I would think they would appreciate being a part of the development of these environmentally sound trains.

Ah yes, the environment. The more our states pave over precious soil, the more problems that creates for the environment. For example, when it rains the water has no place to go, and so it rises and causes flooding. While steps have been taken to reduce the emissions found in automobiles, there is still a sizeable chunk of pollutants that come out of them. Passenger trains may never reduce the flow of auto traffic, but they will surely be a friend of the environment.

Often, we are told that stoplights are timed to have us spend more money on gasoline. Whether this is true or not, whenever a car is stopped or idling it does consume gas and it adds to the problems we face with ozone build up. This alone is a great argument for including intercity and regional trains in the transportation vision. If any of us find that hard to believe, try this. If we live in an urban environment and then look to the heavens at night, how many stars can we see? Almost none. They are there, but air pollutants are one of the reasons why we cannot see them. Another is all the lights that are on in this area.

Are we there yet? In colonial times, the modes of transportation were walking, animal power, and waterways. All modes were slow and, at times, dangerous and expensive. Then in the early 1800's rail was introduced and by the 1900's eliminated the old ways of travel. The trains became an all-weather transport and became the backbone of industrial development. Many of America's great cities, like Chicago, were built on rail and became the financial centers of the nation.

In the 1930's, the automobile became part of the transportation vision and it took away the need for short distance travel by rail. Gradually, station stops began to disappear. It almost took away the need for interurban rail, but somehow trains like the South Shore in Northern Indiana survived the competition.

Air travel came into vogue in the 1950's and began to erode the market for long distance travel to where in 1970 rail travel became minimal. Today with regulations and travel demands, passenger rail is looking attractive again. The airlines are opting out of the short-haul flights for longer distant ones, and the cost of driving a car any length of distance is costly. For many travelers, "getting there" is no longer "half the fun," and they want to leave the mode of travel as quickly as possible.

Passenger rail is changing that. Operators of the systems want the trains to be a pleasurable event and in doing so, are offering amenities that all travelers can enjoy. With all of these positives in favor of high-speed rail, what is keeping the trains from going to a newer, more attractive level?

Roger Sims, former president of the Indiana High Speed Rail Association, suggested the following steps: Advocates need to develop a business model that clearly demonstrates the value of higher-speed trains. They should also look for synergistic relationships with state and regional business groups like the various chambers and business development organizations. Sims also recommended ties with other allies such as the building trades, labor organizations, and leaders in education (Indiana High Speed Rail Association, 2015). I would also suggest that we engage and include students at the various universities.

Rail Magazine, a trade publication, routinely points out that the cost of building passenger rail systems should be seen as an investment, an investment that has measurable returns (Clinnick, 2015).Why would any state want to build more highways that cost far more than a rail corridor, when the demand for greater travel expediency and efficiency is so great?

Cars and airplanes will always be with us, but passenger rail also needs to be an option for travel, and it will be so as long as advocates like National Association of Rail Passengers and the Indiana Passenger Rail Alliance have the heart and voice to demonstrate the need for such travel and the convenience it brings to the travelers. There are still new industries to be built and even newer ways to conduct business. When passenger rail is implemented and becomes a formidable part of the transportation system, there will be greater options for the tourist and leisure traveler, as well.

Europe, Japan, China, and Korea are already there and demonstrating that high-speed travel has a significant place in their societies. Mexico and Brazil will soon launch their high-speed trains to brighten the day of travel in those nations. Indiana, the Midwest, and the nation need to "come on board" now, and fill a serious void that exists in our transportation development.

The rationale behind our proposal is that Indiana and the Midwest need a new approach to public transportation. Our existing systems have reached capacity; and while our awareness of the risks and current course of action is not always clear, the challenges become apparent. As an example, as the population of the United States continues to increase and road and airport congestion consistently interfere with travel plans, the need for passenger rail grows exponentially. Over the next 25 years, the U.S. Census Bureau projects a population increase of 70 million people. That increase places an increased demand on our transportation systems that they currently cannot manage (National Rail Plan Progress Report, 2010). In addition, the U.S. Department of Transportation has issued a series of images showing peak period travel congestion and the projected congestion for 2035 is ominous especially for the state of Indiana and the Midwest (Bureau of Statistics, 2015). Further, utomobile traffic contributes significantly to an enlarged carbon footprint, higher levels of ozone, and the inexorable demand for expanded roadway systems and parking lots. With each passing year, additional species of plants and animals are added to the list of those endangered-or extinct. Greenspace is fast retreating as farmlands become new highways to alleviate congestion, and Ozone Action days, on which we are encouraged to stay indoors and minimize driving for the health of our environment and ourselves, are increasing in number (National Rail Plan Progress Report, 2010).

In terms of global competitiveness and transportation access, as previously stated, congestion in our airports and on our roadways continues to worsen, and projections for future decades are dire. The United States lacks the ease of movement found in most other developed nations. These countries are offering local and long distance passenger rail; and along with air travel and motor vehicles, they are providing alternative means to move from place to place. As global companies weigh their options, America's disconnected, limited transit systems and congested cities are looking less attractive, and existing businesses looking to grow and expand in a global economy are faced with relocation challengers. In addition, airline passengers are faced with new fees and surcharges on top of the already frustrating security procedures and requirements. Once-expected services provided in the past as part of long-distance travel have now become menu options available at additional cost.

Further, some population segments lack good transportation options. Higher-speed or modern 21st Century passenger trains would offer enhanced mobility options to those who do not own or have ready access to an automobile--such as many college students, foreign guests, and some business travelers--or those who may not be able to drive safely, such as some elderly or disabled persons.

According to a report commissioned by the Federal Aviation Administration (FAA) and presented in 2010, domestic flight delays have a $32.9 billion dollar annual cost for the U.S. economy. That number does not address the issue of crowded roadways, which accounted for 4.2 billion hours of lost time in 2007, an average of 36 hours per person, costing the United States an estimated $87.2 billion for wasted fuel and lost productive time (Texas Transportation Institute, 2009).

In a study done by America 2050 and reported by Schned (2014), 27,000 city pairs were analyzed to determine their potential for higher speed rail ridership based on such factors as population size, economic activity, and existing travel markets and transit connections. Among their top 50 city pairs are numerous pairs that would require rail connections through Indiana: (City pairs and likely demand ranking)

- Chicago/Detroit (11)
- Chicago/Columbus (13)
- Chicago/Cleveland (16)
- Chicago/Louisville (35)
- Chicago/Cincinnati (39)

Also, in a cost-benefit analysis done for the Midwest Regional Rail Initiative (MWRRI) by Transportation Economics and Management Systems, Inc. (TEMS) in late 2006, the benefit-cost ratio was found to be 1.8, making the Midwest region a prime area for higher-speed rail--one of the best in the nation (TEMS, 2006).

Further, rail is becoming an increasingly popular form of transit; its popularity would likely increase more with expanded service. Information from the Bureau of Transportation Statistics shows that Amtrak ridership is up about 33% in the past 10 years, from roughly 1.8 million passengers monthly to 2.4 million. Airline traffic usage increased only about 9% during that same time (Bureau of Statistics, 2015).

Another benefit of high-speed rail would be safe and reliable travel. Well-executed higher speed rail will offer reliable travel times. That is, travel time for higher speed rail is much less vulnerable to variable factors such as weather and traffic. For example, rail is safer than car travel, and it is less susceptible to severe weather than air travel. In 2009, U.S. air carriers reported 52 fatalities, 3 train passengers lost their lives, and 23,382 people were killed in highway incidents involving passenger cars or light trucks (Bureau of Statistics, 2015).

According to an article published on Forbes.com, bad weather and poor runway design give Chicago the distinction of having two airports in the top 12 list for delays and cancellations--O'Hare and the much smaller Midway International (Ruiz, 2008). Indianapolis could provide a safer, lake-effect-free alternative destination with easy access to Chicago via high-speed rail.

Higher speed rail can also help to reduce the increasing congestion in our nation's airports and on our nation's highways. According to a Federal Railroad Administration report to Congress in September 2010, these conveyances are largely at capacity and cannot be expanded without extreme difficulty and at great cost.

In some cities, there is simply no room to expand airports or highways, regardless of whether financing is available to accomplish these projects. A high-speed and intercity passenger rail network will complement existing transportation choices and, in many cases, relieve some of the increased demand.

On the other hand, the 110 mph version of high-speed rail utilizing existing rights-of-way requires acquisition of very little extra land, thereby minimally disrupting communities and nature. Further, higher speed rail will increase travelers' productive time, allowing them to utilize electronic devices at any time during travel for work or enjoyment, while also allowing for activities such as conferences, relaxation, and freedom of movement.

Expanding our options by building more energy efficient modes of public transportation makes sense from an energy stewardship perspective, from a supply-disruption risk perspective, and from a transportation cost-containment perspective. Further, in the comparative study of mass transit, rail is shown to use the least energy, measured in BTUs, per passenger mile Auto: 3,437; Air Carrier: 3,051; Commuter Train: 2,656; and Intercity Train: 2,398 (National Plan Progress Report, 2010). Rising oil prices adversely affect nearly everyone in America, taking up a greater share of household and national budgets. With increasing oil demand from other developed and developing nations, the cost is likely to rise even faster in coming years. As a commuter society, the U.S. will need transportation alternatives to keep individuals and the economy on the move.

Taking a long-term perspective, petroleum is a finite resource. Over the next 50 to 100 years, energy prices are likely to rise significantly. The public tends to focus only on short term energy price movements, and continues to seek ways to sidestep rising costs primarily by looking for ways to do more driving with less fuel. This short-term focus distracts us from consideration of long-term risks. We need to be actively exploring and creating other transportation options that reduce our national dependence on petroleum and provide ways to reduce congestion and improve travel safety in the long run – and high-speed rail is certainly a viable option!

So are we there yet? Not yet, but we should certainly get there soon. Most of the countries in the world around us, especially in Europe, Asia, South America already have sophisticated, thriving, growing, and efficient high-speed rail systems and America needs to quickly catch up. We certainly can do so as we see more high-speed rail developments potentially in California, Florida, Nevada, Texas, Midwest including Illinois, Indiana, and other travel corridors.

REFERENCES

Clinnick, R. (2015). Investment is changing perception of reality. *RailMagazine.com*. Retrieved from http://www.railmagazine.com/news/network/2015/11/18/investment-is-changing-perception-of-rail-perry

Department of Transportation, Bureau of Statistics. (2015). Retrieved from http://www.rita.dot.gov/bts/home

Florida, R. (2009). How the crash will reshape America. *The Atlantic*. Retrieved from http://www.theatlantic.com/magazine/archive/2009/03/how-the-crash-will-reshape-america/307293/

Indiana High Speed Rail Association. (2015). Retrieved from http://www.indiana-highspeedrail.org/

National Rail Plan Progress Report. (2010). National Rail Plan. Retrieved from https://www.fra.dot.gov/eLib/Details/L02696

Ruiz, R. (2008). America's most time-draining airports. *Forbes.com.* Retrieved from http://www.forbes.com/2008/06/02/aviation-travel-delays-biz-logistics-cx_rr_0603travel.html

Schned, D. (2014). FRA: California High-Speed Rail Phase 1 Clear for Construction. Retrieved from http://www.america2050.org/

SIEMENS. (2010). Building High Speed Rail in America. Retrieved from http://www.coaststarlight.net/railpac/2010steelwheelsinca/Armin%20Kick%20-%204.17.10%20Steel%20Wheels%20in%20CA.pdfSIEMENS

TEMS. (2006). Midwest Regional Rail Initiative Benefit Cost & Economic Analysis http://www.iowadot.gov/iowarail/passenger/mwrri-economic.pdf

Texas Transportation Institute. (2009). Urban Mobility Report. Retrieved from http://d2dtl5nnlpfr0r.cloudfront.net/tti.tamu.edu/documents/mobility-scorecard-2015.pdf

Chapter 13
High Speed Rail:
Suggestions for Case Studies, Research Questions, and Projects

Raj Selladurai
Indiana University Northwest, USA

George VandeWerken
Providence Bank, USA

ABSTRACT

High-speed rail is gaining momentum worldwide in many countries in the world including the United States and especially in the state of California, Florida, and Texas currently. Focusing on the list of topics below would enable readers including professionals, policy-makers, leaders, staff, academicians, scholars, and students to explore assignments and research into innovative ideas, plans, and strategies related to high-speed rail in the U.S. The chapter presents a list of topics that includes case studies, research questions, projects, and other suggestions to stimulate further research and explore assignments into more optimal formulation and implementation of high-speed rail for the U.S. in the future.

DOI: 10.4018/978-1-5225-0102-2.ch013

INTRODUCTION

High-speed rail is gaining momentum worldwide, including in the United States -- especially in the states of California, Florida, and Texas currently. Focusing on the list of topics below would enable readers – including professionals, policy-makers, leaders, staff, academicians, scholars, and students – to explore assignments and research innovative ideas, plans, and strategies related to formulation, design and development, and implementation of high-speed rail in the U.S.

LIST OF TOPICS

Case Studies

http://www.allaboardflorida.com
http://www.hsr.ca.gov
http://www.indianahighspeedrail.org/
http://www.midwesthsr.org/
http://niprarail.org/
http://www.txdot.gov/inside-txdot/projects/studies/statewide/dallas-houston-high-speed-rail.html
http://www.ushsr.com/

Research Questions/Areas/Topics to Explore:

1. How may we replicate the California High Speed Rail development and implementation experience in other travel corridors?
2. How do we build continuous support for California high-speed rail within California constituencies?
3. How do we ensure optimal implementation of high-speed rail for California in the future?
4. Which funding model is most efficient for high-speed rail development – public, private, or a blended, hybrid approach?
5. Develop an integrated, multi-modal optimal transportation model customized for a specific region in the U.S.
6. Explore further research into formulation and implementation of an airline and high-speed rail collaboration model.
7. Explore formulation and implementation of high-speed rail in the Southwest region of the U.S.

8. Given the map (Figure 1) of a high-speed rail vision (http://www.ushsr.com/), how do we more effectively implement this vision into reality in the future? What strategies may be used to facilitate the formulation and implementation of this vision into reality --in specific travel corridors? In specific time periods/ phases?

Example of a Potential Research Project

Exploring further into formulation and implementation of an airline and high-speed rail collaboration model. See Chapter 7.

The airline/high-speed rail model is premised on airline control or "ownership" of the passenger travel experience from the point of origin to the point of destination, regardless of the infrastructure used to move passengers and baggage. This is a departure from the standard model currently used. Airlines now move passengers exclusively on planes. Another feature of the model is a revised role for Amtrak. Currently, Amtrak owns most of the rolling stock and operates most of the trains in this country. The proposed model would assign Amtrak with the task of negotiating for and developing the infrastructure and then contacting with private parties, such

Figure 1. US HSR network phasing plan
(http:www.ushsr.com)

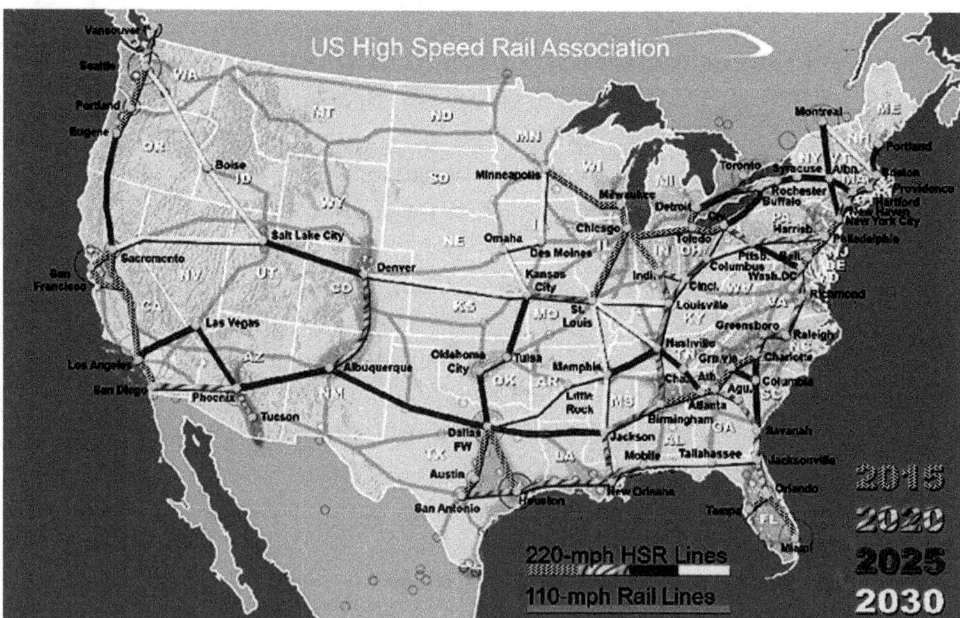

as the airlines, to operate service on these lines. This is analogous to the current relationship between airline carries and municipalities – municipalities own airports but they don't fly planes. Amtrak might provide the infrastructure on which others are licensed to operate.

In order for the new model to work, however, a number of infrastructure enhancements need to be put in place. A research project to explore the feasibility of this new suggested model could include the following steps:

1. Determine the current cost to move passengers and baggage between and two city pairs using air travel. In our example, we used Indianapolis and Chicago, but other city pair examples could be explored. It is important to know current airline costs for city pairs because airlines will not be interested in an alternative model if it increases their cost.
2. Determine the projected cost to move passengers and baggage between selected city pairs using the airline train travel model. To do this requires an estimate of the cost to build and maintain the required infrastructure, and to operate on the new system. This includes at least the following:
    a. Train "gates" at the airports to allow passengers to move within the airport from plane to train or train to plane. Since passengers would go through airport security to board the plane or train, the train "gates" have to be designed to board passengers behind the TSA security system.
    b. Baggage handling integration for the train "gates" so that baggage can be moved between trains, planes and carousels.
    c. Rail line construction or right of way acquisition and improvements. The real estate on which the rail lines operate is owned primarily by the freight railroads. By law, freight rail operators only need to negotiate with Amtrak for passenger service. So Amtrak will need to be engaged. Further, in order for the model to work, the right of way must be designed and built to handle trains at speeds of 125 miles per hour or better. This will require significant enhancements to existing infrastructure, not only for passenger service, but to avoid interference with freight service running in the same rail corridors. One significant and costly infrastructure component to prevent collisions with road vehicles is the elimination of rail grade crossings for the higher train speeds. Another expensive component is positive train control systems. These are only two examples of cost estimates that need to be developed.
    d. Rolling stock acquisition. Sufficient rolling stock must be acquired to accommodate the desired frequency of service. For example, if half hour service frequency is desired from each endpoint, and a one-way trip takes two hours, then at least eight train sets must be acquired to provide enough

rolling stock to achieve that level of service. If quarter hour service is desired, then at least sixteen train sets must be acquired and operated.

3. Determine annual infrastructure maintenance costs. Once the cost to build or acquire the requisite infrastructure is determined, the fixed and variable cost elements need to be factored in. The fixed infrastructure costs include financing costs, depreciation, and maintenance. The variable costs include train set operation and revenue sharing arrangements with Amtrak or the freight railroads.
4. Determine passenger levels. Once projected costs are determined, an estimate of the required passenger levels needed in order to generate positive cash flow and profitable operation can be determined. That number can be compared to existing airline passenger traffic to determine how many additional passengers need to be attracted to generate profitable operations.
5. Evaluate costs and impact of intermediary stops. If trains make stops at locations other than airports, the challenge is how to maintain TSA security standards for any passengers coming onto the train. Should the train be designed with TSA secure cars and others that are not? Are some train departures exclusively TSA compliant, allowing no intermediate onboarding, and other departures non TSA compliant, requiring security screening once the passengers arrive at the airport? Should intermediary train stations begin handling TSA security? All these options present different operational considerations and costs.
6. Determine other revenue enhancement opportunities. For example, is there is a parcel handling component that could be added to generate additional revenue? Would UPS or FedEx be interested in moving some of their parcels on these trains? Are there other revenue generating amenities that could be added? What other passengers might be attracted to this service who now prefer to drive or who would be served by intermediary stops?
7. Evaluate operating models. Amtrak could own and operate the power units (engines) and provide the train crews, and airlines could provide the passenger cars and passenger car staff. A train could consist of eight to ten cars, but several airlines could have cars on the same train, or one airline might take all cars on the train exclusively. If some cars are intended not to be TSA compliant, the airlines or other parties could operate cars that would accommodate travelers who want to travel from one city to another with actually going though security into the air terminal building. Obviously, each option offers different costs and different revenue opportunities.
8. Compare current airline costs for passenger movement to projected costs to move those passengers by train. The study should provide a comparison of the options within the model and a breakeven analysis as well as a best and worst

case estimate. One of the tasks of the study is to develop realistic estimates of passenger service expansion that could be realized with new, enhanced types of service.

These ideas are intended to illustrate at least some of the elements that could be included in a research project. This outline is not exhaustive, and as the project gets underway, additional considerations will become evident. Because of the size and complexity of a project like this, it may be prudent to break the project into separate sub-projects that could ultimately be assembled into a final presentation. Breaking the project down also allows groups to work with different interested or affected constituencies -- airlines, for example, or airport terminal operators, or Amtrak, or equipment manufacturers, or other consultants, scholars, and professors who can help with various aspects of the overall project.

Compilation of References

American Public Transportation Association. (2012, May). APTA High-Speed Train Survey.

AASHTO Journal. (2015). *Update: INDOT extends 'Hoosier State' passenger train support through April.* Retrieved from http://www.aashtojournal.org/Pages/031315Hoosier.aspx)

Airports Council International/North America. (2011). Airport capital development costs 2011 – 2015.

Albalate, D., & Bel, G. (2012). High-speed rail: Lessons for policy makers from experiences abroad. *Public Administration Review*, *72*(3), 336–349. doi:10.1111/j.1540-6210.2011.02492.x

Alberta Association of Municipal Districts and Counties. (2010, October 1). *Study of high speed rail impacts on rural Alberta, final report.* Retrieved from http://www.aamdc.com/component/docman/doc_details/78-2010-study-of-high-speed-rail-impacts-on-rural-alberta?Itemid=208

Alstom. (2009). *AGV: Full-Speed Ahead Into the 21st Century.*

Alstom. (n. d.). *The AGV, a New and Revolutionary Very High Speed Train.*

American Association of State Highway and Transportation Officials. (2014). Retrieved from www.highSpeed_rail.org/pages/basicfacts.aspx

American Lung Association (ALA). (2013). Most Polluted Cities. Retrieved from http://www.stateoftheair.org/2013/city-rankings/most-polluted-cities.html

American Public Transportation Association. (2012). 10.4 billion trips taken on U.S. public transportation in 2011.

American Public Transportation Association. (2014). Record 10.7 Billion Trips Taken On U.S. Public Transportation In 2013.

American Road & Transportation Builders Association. (2014). Transportation FAQs. Retrieved from www.artba.org/about/transportation-faqs/#20

American Society of Civil Engineers. (2014). 2013 report card for America's infrastructure. Retrieved from www.infrasttructurereportcard.org

Amos, P., Bullock, D., & Sondhi, J. (2010). *High-speed rail: the fast track to economic development, World Bank*. Beijing: R.P., China.

Amtrak website. (2011). Atlanta Constitution. (March 31, 2006). On-time arrival for fifth runway at Atlanta Hartsfield-Jackson Int'l Airport.

Amtrak website. (2014). *Long distance train facts.*

Amtrak. (2012). *Amtrak on pace to set new ridership records.*

Amtrak. (2014). *FY 2015 budget request seeks change in federal passenger rail investment: warns status quo puts Northeast Corridor infrastructure at risk.*

Australian Government, Department of Infrastructure, Transport, Regional Development and Local Government. (2010). *A profile of high-speed railways.*

Avril, E. (2010). *Articulated/Conventional Architecture. In 2010.* La Rochelle, France: William Barclay Parsons Fellowship.

Balakrishnan, J., Eliasson, J. B., & Sweet, T. R. C. (2007). Factors affecting the evolution of manufacturing in Canada: An historical perspective. *Journal of Operations Management*, *25*(2), 260–283. doi:10.1016/j.jom.2006.05.014

Ball, M., Barnhart, C., Dresner, M., Hansen, M., Neels, K., Odoni, A., & Zou, B. et al. (2010). *Total delay impact study: a comprehensive assessment of the costs and impacts of flight delay in the United States final report.* National Center of Excellence for Aviation Operations Research.

Banko, F., & Xue, J. (2012). *Pioneering the Application of High Speed Rail Express Trainsets in the United States (50).* New York, NY: Parsons Brinckerhoff Group, Inc.

Baxandall P., Davis B., & Dutzik T. (2012). *Transportation and the new generation: why young people are driving less and what it means for transportation policy.* Frontier Group/U.S. PIRG Education Fund.

Baxandall, P. (2013, August). *Moving off the Road: A State-by-State Analysis of the National Decline in Driving.* U.S. PIRG Education Fund.

Berlioz, C., & Leboeuf, M. (1986). Performance of the Paris-southeast TGV. *Revue General des Chemins de Fer*, *196*, 759–768.

Bernard, A. B., Moxnes, A., & Saito, Y. U. (2014, September 24). Fast trains, supply networks, and firm performance. Centre for Economic Policy Research, UK. Retrieved from http://www.voxeu.org/article/fast-trains-and-firm-performance-new-japanese-evidence

Bloomberg Business. (2015). *Company overview of Iowa Pacific holdings, LLC.* Retrieved from http://www.bloomberg.com/research/stocks/private/snapshot.asp?privcapId=13500461

BNSF Railway. (2014). Safety and Security. Retrieved from http://www.bnsf.com/communities/safety-and-security/railroad-grade-crossings/

Bureau of Transportation Statistics, Research and Innovative Technology Administration. (2011). *Annual report on airline performance*. Washington, D.C.: U.S. Department of Transportation.

California Department of Finance. (n. d.). Report P-1: Summary Population Projections by Race/Ethnicity and by Major Age Groups. Retrieved from http://www.dof.ca.gov/research/demographic/reports/projections/P-1/

California High-Speed Rail Authority. (n. d.) Retrieved from http://www.hsr.ca.gov/

Canadian Broadcasting Corporation. (2014, April 15). Liberals promise high speed rail in Kitchener within 10 years. Rail line would link Toronto, Pearson, Kitchener and London. Retrieved from http://www.cbc.ca/news/canada/kitchener-waterloo/liberals-promise-high-speed-rail-in-kitchener-within-10-years-1.2610303

Central, J. R. (2015, February 1). Retrieved from http://english.jr-central.co.jp/company/company/achievement/financeandtransportation/transportation2.html

Clinnick, R. (2015). Investment is changing perception of reality. *RailMagazine.com.* Retrieved from http://www.railmagazine.com/news/network/2015/11/18/investment-is-changing-perception-of-rail-perry

Craven, P., & Traves, T. (1983). Canadian railways as manufacturers, 1850-1880. *Historical Papers*, *18*(1), 254–281.

Cuny, N. America Referential for High Speed. In 2010William Barclay Parsons Fellowship. La Rochelle, France.

De Rus, G. (2008). The economic effects of high-speed rail investment: discussion paper no. 2008-16 revised. Joint Transport Research Centre, University of Las Palmas, Spain.

de Rus, G. (Ed.), (2009). *Economic analysis of high speed rail in Europe.* Bilbao Spain: Fundación BBVA. Retrieved from http://www.fbbva.es/TLFU/dat/inf_web_economic_analysis.pdf

Department of Transportation, Bureau of Statistics. (2015). Retrieved from http://www.rita.dot.gov/bts/home

Dumaine, B. (2015, June 4). Why America is betting big on bullet trains. *Fortune.com.* Retrieved from http://fortune.com/2015/06/04/americas-bet-on-bullet-trains/

Dutzik, T., & Baxandall, P. (2013). A New Direction Our Changing Relationship with Driving and the Implications for America's Future. U.S. PIRG Education Fund, Frontier Group.

Dutzik, T., & Steva, E. (2010). *Next stop California: the benefits of high-speed rail around the world and what's in store for California.* California Public Interest Research Group Education Fund.

Dykstra, M. (2014, May 23). Edmonton and Calgary should think about high-speed rail system, but shouldn't put money into it. *Edmonton Sun.* Retrieved from http://www.edmontonsun.com/2014/05/23/edmonton-and-calgary-should-think-about-high-speed-rail-system-but-shouldnt-put-money-into-it

EcoTrain. (2011). *Updated feasibility study of a high speed rail service in the Quebec City – Windsor corridor*. Montreal, Quebec: EcoTrain.

Elking, I., & Windle, R. (2014, Fall). Examining differences in short-haul and long-haul markets in US commercial airline passenger demand. *Transportation Journal*, *53*(4), 424–452. doi:10.5325/transportationj.53.4.0424

ENO Transportation Foundation. (2012). *NextGen: aligning costs, benefits and political leadership*. Washington, D.C.

European Trains. (n. d.). *Rail Europe*. Retrieved from https://www.raileurope.com/european-trains/

Federal Aviation Administration. (2014), *Fact sheet – FAA forecast – fiscal year 2013 – 2014*. Retrieved from www.gov/news/fact_sheet/news_story.cfm?newsld=15934

Federal Highway Administration. (2009). *National household travel survey*. Washington, D.C.: U.S. Department of Transportation.

Federal Highway Administration. (2013). Status of the nation's highways, bridges, and transit: conditions & performance. *U.S. Department of Transportation,* Washington, D.C., www.fhwa.dot.gov/policy/2013cpr/lb_es.htm

Federal Railroad Administration. (2009). *High-Speed Passenger Rail Safety Strategy*, 1-2.

Federal Railroad Administration. (2009). *Vision for High-Speed Rail in America*, *1*, 6.

Federal Railroad Administration. (2010). *RSAC Engineering Task Force*. Cambridge, MA.

Federal Railroad Administration. (2013). *Midwest Regional Investment Fact Sheet*. Washington, D.C.: U.S. Department of Transportation. Retrieved from http://www.fra.dot.gov/eLib/details/L04666

Federal Railroad Administration. (2013). Summary. In 49 CFR Parts 213 and 238 Vehicle/Track Interaction Safety Standards; High-Speed and High Cant Deficiency Operations; Final Rule, 16052.

Federal Railroad Administration. (2014, September). *Southwest multi-state rail planning study: Summary report,* 1-11. CrossRailChicago: A vision for a modern Metra. Retrieved from http://CrossRailChicago.org

Federal Railroad Administration. (2015, August). *Dallas to Houston High-Speed Rail Project Corridor Alternatives Analysis Technical Report*. Retrieved from https://www.fra.dot.gov/eLib/details/L16978

Florida Department of Transportation. (2011, April 15) Airport costs. Retrieved from http//www.Dot.state.fl.us/planning/policy/cost

Florida, R. (2009). How the crash will reshape America. *The Atlantic*. Retrieved from http://www.theatlantic.com/magazine/archive/2009/03/how-the-crash-will-reshape-america/307293/

Global Air Rail Awards. (2012). *Case Studies: Intermodal Partnerships*. Retrieved from https://www.globalairrail.com/images/content/publications/casestudies/Winners_Case_Studies_Partnerships.pdf

Gopnik, A. (2015). *The plot against trains*. Retrieved from http://www.newyorker.com/news/daily-comment/the-plot-against-trains

Government Accountability Office. (2009). *High-speed passenger rail: future development will depend on addressing financial and other challenges and establishing a clear federal role, GAO-09-317.*

Government of Alberta. (2013, September 26), *Alberta's population reaches four million.* Retrieved from http://alberta.ca/release.cfm?xID=350618FCB7CB9-D315-C7B6-3A1E8521B366D28A

Government of Alberta. (2014, September 30). Electricity facts. Retrieved from http://www.energy.alberta.ca/Electricity/681.asp

Greengauge, (2006), *High-speed trains and the development and regeneration of cities.*

Greenlaw, C. C. N. (2007). *VIA rail.* St. Paul, MN: Voyageur Press.

Greyhound. (2016). Retrieved from https://greyhound.com/Express/Default.aspx

Haas, P. (2014). Modal shift and high-speed rail: a review of the current literature. *Mineta Transportation Institute, Report 12-35.*

Hagler, Y., & Todorovich, P. (2013). *Where High-Speed Rail Works*, New York: American 2050. Retrieved from http://www.america2050.org/pdf/Where-HSR-Works-Best.pdf

Hara, M., Kurita, T., Mizushima, F., Satoh, H., Shikama, T., Wakabayashi, Y., & Yamada, H. (2009). Reduction of Pantograph Noise of High-Speed Trains. *Journal of Mechanical Systems for Transportation and Logistics*, *3*(1), 73.

Haynes, K. E., & Fotheringham, A. S. (1984). *Gravity and Spatial Interaction Models*. Beverly Hills: Sage Publications.

Henry, L., & Marsh, D. (2009), Intermodal surface public transport hubs: harnessing synergy for success in America's urban and intercity travel, *Capital Metropolitan Transportation Authority*, Austin, Texas. Retrieved from http://www.amtrak.com/servlet/ContentServer/Page/1248542787937/1237405732517

High-speed rail in Germany. (n. d.). Wikipedia. Retrieved from https://en.wikipedia.org/wiki/High-speed_rail_in_Germany

Hong Kong Economic Journal. (2016). *Airlines explore air rail business collaboration*. Retrieved from http://www.ejinsight.com/20140113-airlines-explore-air-rail-business-collaboration-paper-says/

http://www.progressiverailroading.com/high_speed_rail/article/Midwestern-states-hope- to-take-their-longplanned-highspeed-and-intercity-passenger-rail-network-to-the- next-level--20806

Hu, M., Patel, A., Simmons, J., & Tsu, J. (2014, April 29). Feasibility of the implementation of a high-speed rail system – Phasing out regional air travel. *AOSS 480 – Final Report.*

Hughes, M. (2015). *The Second Age of Rail: A History of High-Speed Trains*. London, UK: The History Press.

Hurley, W. J., & Jones, J. (1990). *The feasibility of high-speed rail in Canada*. Kingston, Ontario: Department of Economics, Queen's University.

Indiana Department of Transportation (INDOT). (2015). Retrieved from http://www.in.gov/indot/

Indiana Department of Transportation. (2015). Average daily traffic and commercial vehicles interactive map. Retrieved from https://entapps.indot.in.gov/TrafficCounts/

Indiana High Speed Rail Association. (2015). Retrieved from http://www.indianahighspeedrail.org/

Indiana, High Speed Rail: 'LINK' to the Future. (2013). *The South Shore Journal*, *5*, 171-182.

Indianapolis International Airport. (2016). Retrieved from http://ind.fltmaps.com/en

International Union of Railways (UIC). (2014). International level Crossing Awareness Day (ILCAD). Retrieved from http://www.ilcad.org/History.html

International Union of Railways (UIC). (2014). RESTRAIL. Seventh Framework Programme for Research and Technological Development. Retrieved from http://www.restrail.eu/

International Union of Railways (UIC). (2014). Safety Database Report Significant Accidents occurred in Europe during the year 2013 Public Report [Data file]. Retrieved from http://safetydb.uic.org/IMG/pdf/Public_Report_2014.pdf

International Union of Railways. (2011, July 27). UIC world statistics. Retrieved from http://www.uic.org/com/IMG/pdf/cp18_uic_stats_2010_en.pdf

International Union of Railways. (2014, September 1). High speed lines in the world. Retrieved from http://www.uic.org/IMG/pdf/20140901_high_speed_lines_in_the_world.pdf

Jeanes, D. (2009, May 28). *Meeting of the Standing Committee on Transport, Infrastructure and Communities.* Parliament of Canada, Ottawa, Ontario. Retrieved from www.parl.gc.ca/House-Publications/Publication.aspx?DocId=3935511&Language=E&Mode=2&Parl=40&Ses=2

Jehanno, A., James, C., & Palmer, D. (2011). *High-speed rail and sustainability*. International Union of Railways.

Joint Transportation Research Centre. (2008). *Effects of high-speed rail investment discussion paper no. 2008-16.*

Kageson, P. (2009). Environmental aspects of inter-city passenger transport (Discussion paper no. 2009-28). *Joint Transport Research Centre.*

Kanter, M. (2015). *Move: Putting America's infrastructure back in the lead.* London, UK: W.W. Norton & Company. Retrieved from http://www.amazon.co.uk/Move-Putting-Americas-Infrastructure-Back/dp/0393246809

Kelly, A. (2016). Maglev Train: USA gets $28M study for 300 mph train travel that gets from D.C. to NYC in less than an hour. *IDigitalTimes.com.* Retrieved from http://www.idigitaltimes.com/maglev-train-usa-gets-28m-study-300-mph-train-travel-gets-dc-nyc-less-hour-490793

Kelly, B., & Nichols, M. (2014, January 29). Taking California's bullet train to a greener future. *LATimes.com.* Retrieved from http://www.latimes.com/opinion/op-ed/la-oe-kelly-california-high-speed-rail-20140129-story.html

Knowlton, B. (2009). Obama Seeks High-Speed Rail System Across U.S. *New York Times.* Retrieved from http://www.nytimes.com/2009/04/17/us/politics/17train.html?_r=0

Koschnitzke, A., Schüler, J., & Waldeck, H. (2010). *Acoustics. In 2010.* Krefeld-Uerdingen, Germany: William Barclay Parsons Fellowship.

Kurita, T. & Mizushima, F. (n. d.). Environmental Measures along Shinkansen Lines with FASTECH360 High-Speed Test Trains. *JR East Technical Review, 16*, 48, 49, 51, 53.

Lacôte, F. (2010). *Very High Speed Train and Safety. In 2010.* Paris, France: William Barclay Parsons Fellowship.

Langer, A., & Winston, C. (2006). *The effect of government highway spending on road users' congestion costs.*

Le Corre, D. (2010). *Passive Safety in Alstom High Speed Trains. In 2010.* Reichshoffen, France: William Barclay Parsons Fellowship.

Li, Y. (2010). *Information about High Speed EMU. In 2010.* Qingdao, China: William Barclay Parsons Fellowship.

Machalaba, D. (2007). Crowds heed Amtrak's 'all aboard,' improved service, air, woes lure travelers in northeast; long hauls still suffer. T*he Wall Street Journal.*

Massachusetts Institute of Technology. (2015). *Global airline industry program: Airline industry overview.* Retrieved from http://web.mit.edu/airlines/analysis/analysis_airline_industry.html

MegaBus. (2016). Retrieved from http://us.megabus.com/Default.aspx

Metropolitan Washington Airports Authority. (n. d.). Retrieved from (www.washingtonairports.com/dulles/773

Midwest High Speed Rail Association. (2011). *The Economics of Impacts of High Speed Rail: Transforming the Midwest*. Midwest High Speed Rail Association: Chicago, IL. Retrieved from www.midwesthsr.org

Midwest High Speed Rail Association. (n. d.). Retrieved from http://www.midwesthsr.org/

Midwest Interstate Passenger Rail Commission. (2000). *Midwest Regional Rail System: Transportation System for the 21st Century*. Lombard, IL: Midwest Interstate Passenger Rail Commission.

Midwest Interstate Passenger Rail Commission. (2004). *Midwest Regional Rail Initiative Report*. Lombard, IL: Midwest Interstate Passenger Rail Commission.

Milmo, D. (2012). Virgin empire: from selling records to running trains and airline. *The Guardian*. Retrieved from http://www.theguardian.com/business/2012/aug/15/virgin-train-richard-branson

Mineta Transportation Institute. (n. d.). Retrieved from http://transweb.sjsu.edu/

Moijawad, J. (2011), Air service cutbacks hit hardest where recession did. Retrieved from www.newyorktimes.com

Morgan, C., & Sperry, B. (2011). Economic impacts of intercity passenger rail service: evidence from passenger surveys, *Transportation Research Record: Journal of the Transportation Research Board, No. 2261*, Washington, D.C.

Morrison, S., & Winston, C. (2008). The effect of FAA expenditures on air travel delays. *Journal of Urban Economics*, 2008, 63.

Murphy, D., & Meilus, A. (2012, January). Measurement of normalized change in demand for short-haul O&D commercial air rravel from 1995 to 2010. *Paper presented atTransportation Research Board 91st annual meeting*, Washington, DC.

Nash, C. (2009). When to invest in high-speed rail links and networks. *OECD/International Transport Forum*, Madrid, Spain.

Nash, C. (n. d.). High Speed Rail Investment: an overview of the literature. *Institute for Transport Studies, University of Leeds.*

National Council on Strength and Fitness. (n. d.). Obesity Awareness Month. Retrieved from http://www.ncsf.org/enew/articles/articles-obesityawarenessmonth.aspx

National Rail Plan Progress Report. (2010). National Rail Plan. Retrieved from https://www.fra.dot.gov/eLib/Details/L02696

National Surface Transportation Infrastructure Financing Commission. (2008). *Paying Our Way: A New Framework for Transportation*. Washington, D.C.: U.S. Department of Transportation.

Neff, J. (2011). *High-speed rail investment background data*. Washington, D.C.: American Public Transportation Association.

Nixon, R. (2012). *Frustrations of air travel push passengers to Amtrak.* Retrieved from http://www.nytimes.com/2012/08/16/business/hassles-of-air-travel-push-passengers-to-amtrak.html?_r=0

Obama, B. (2011). *State of the Union Address.* Washington, D.C.

Operation Lifesaver. (2014). Mission Statement. Retrieved from http://oli.org/about-us

Organization for Economic Cooperation and Development. (2001), http://esa.un.org/unsd/sna1993/introduction.asp

Paddock, G., & Hayhurst, T. (2009, March 18). Northeast Indiana Passenger Rail Association (NIPRA) is formed guest. Retrieved from http://www.thewaynedalenews.com

Passenger Rail Working Group. (2007). *Vision for the future U.S. intercity passenger rail network through 2050.* National Surface Transportation Policy and Revenue Study Commission.

Pepy, G. & Perren, B. (2006). 25 years of the TGV. *Modern Railways,* October, 67- 74.

Pomfret, R. (1981). *The economic development of Canada.* Toronto, Ontario: Methuen.

Reno A. & Weisbrod G., (2009). *Economic Impact of Public Transportation Investment.*

Roberts, S. (2013). All Aboard. In *Grand Central – How a Train Station Transformed America (114).* New York, NY: Hachette Book Group.

Ruiz, R. (2008). America's most time-draining airports. *Forbes.com.* Retrieved from http://www.forbes.com/2008/06/02/aviation-travel-delays-biz-logistics-cx_rr_0603travel.html

Santini, D., Wang, M., & Warner, S. (1994). *Methods of valuing air pollution and estimated monetary values of air pollutants in various U.S. regions, Argonne National Laboratory.* Argonne, IL: U.S. Department of Energy.

Savage, I. (2014) Analysis of Fatal Train-Pedestrian Collisions in Metropolitan Chicago 2004-2012.*Proceedings of the 2014 Global Level Crossing and Trespass Prevention Symposium.* University of Illinois at Urbana-Champaign.

Savage, I., & Mok, S. (2005, August). Why has Safety Improved at Rail-Highway Grade Crossings? *Risk Analysis, 25*(4), 867–881. doi:10.1111/j.1539-6924.2005.00642.x PMID:16268935

Schmidt, G., & Waldeck, H. (2010). *Crashworthiness. In 2010.* Krefeld-Uerdingen, Germany: William Barclay Parsons Fellowship.

Schned, D. (2014). FRA: California High-Speed Rail Phase 1 Clear for Construction. Retrieved from http://www.america2050.org/

Schroeder-Bodenstein, K. (2010). *Running Dynamics and Gauging. In 2010.* Krefeld-Uerdingen, Germany: William Barclay Parsons Fellowship.

Schwieterman, J. (2014). *Terminal Town: An Illustrated Guide to Chicago's Airport, Bus Depots, Train Station and Steamship Landings: 1939 - Present*. Lake Forest: Lake Forest College Press.

Schwieterman, J., & Scheidt, J. (2007). Survey of Current High-Speed Rail Planning Efforts in the United States. *Transportation Research Record: Journal of the Transportation Research Board, 1995*, 27–34. doi:10.3141/1995-04

Selladurai, R. (2013). Indiana high-speed rail: 'LINK' to the future. *The South Shore Journal, 5*, 171–182.

Shinkansen Map. (2016, February 8). *Wikimedia*. Retrieved from http://commons.wikimedia.org/wiki/File:Shinkansen_map_201503_en.png

SIEMENS. (2010). Building High Speed Rail in America. Retrieved from http://www.coaststarlight.net/railpac/2010steelwheelsinca/Armin%20Kick%20-%204.17.10%20Steel%20Wheels%20in%20CA.pdfSIEMENS

Sivak, M., & Schoet, B. (2011). *Recent changes in the age composition of drivers in 15 countries UMTRI-2011-43*. University of Michigan Transportation Research Institute.

Smith, K. (2014, May 2). Ontario government backs Toronto - London HSL. *International Railway Journal*. Retrieved from http://www.railjournal.com/index.php/high-speed/ontario-government-backs-toronto-london-hsl.html?channel=542

Southeastern (train operating company). (2016, February 8). *Wikipedia*. Retrieved from https://en.wikipedia.org/wiki/Southeastern_(train_operating_company)

Standing Committee on Alberta's Economic Future of the Legislative Assembly of Alberta. Edmonton. (2014, May). *The feasibility of establishing a high-speed rail transit system in Alberta*. Retrieved from http://s3.documentcloud.org/documents/1173385/high-speed-rail-transit-system-in-alberta-final.pdf

Systra. (2009). *High-speed rail development programme 2008/9, final report for Workstream 2.*

TEMS, Inc. (2004). Midwest Regional Rail Initiative Project Notebook (Ch. 4).

TEMS. (2006). Midwest Regional Rail Initiative Benefit Cost & Economic Analysis http://www.iowadot.gov/iowarail/passenger/mwrri-economic.pdf

Texas Transportation Institute. (2009). Urban Mobility Report. Retrieved from http://d2dtl5nnlpfr0r.cloudfront.net/tti.tamu.edu/documents/mobility-scorecard-2015.pdf

Texas Transportation Institute. (2011). Annual Urban Mobility Report, Texas A&M University.

Transport Canada. (2014). Retrieved from http://www.tc.gc.ca/eng/rail-menu.htm

Transportation for America. (2011), *Aging in Place, Stuck without Options: Fixing the Mobility Crisis Threatening the Baby Boom Generation*. Retrieved from http://t4america.org/resources/seniorsmobilitycrisis

U.S. Census Bureau. (2011, March). Population Distribution and Change: 2000-2010.

U.S. Department of Transportation. (1997). *High-Speed Ground Transportation for America.*

U.S. Department of Transportation. (2012). *2013 Budget.*

U.S. Department of Transportation. Bureau of Transportation Statistics. (2014). National Transportation Statistics, Chapter 2 – Auto Fatalities in 2010 [Data File]. Retrieved from http://www.rita.dot.gov/bts/sites/rita.dot.gov.bts/files/publications/national_transportation_statistics/index.html#chapter_2

U.S. Department of Transportation. Bureau of Transportation Statistics. (2014). National Transportation Statistics, Chapter 2 – General Aviation Fatalities in 2010 [Data file]. Retrieved from http://www.rita.dot.gov/bts/sites/rita.dot.gov.bts/files/publications/national_transportation_statistics/index.html#chapter_2

U.S. Department of Transportation. Bureau of Transportation Statistics. (2014). National Transportation Statistics, Chapter 2 – Transit Fatalities in 2010 [Data file]. Retrieved from http://www.rita.dot.gov/bts/sites/rita.dot.gov.bts/files/publications/national_transportation_statistics/index.html#chapter_2

U.S. Department of Transportation. Federal Railroad Administration – Office of Safety. (2009) FRA High-Speed Passenger Rail Safety Strategy; Retrieved from https://www.fra.dot.gov/eLib/Details/L03624

U.S. Department of Transportation. Federal Railroad Administration – Office of Safety. (2014). Retrieved from http://www.fra.dot.gov/Page/P0010

U.S. Department of Transportation. Federal Railroad Administration. (2014) 49 CFR Ch. II Part 209 Appendix A. Statement of Agency Policy Concerning Enforcement of the Federal Railroad Safety Laws. Washington, DC: U.S. Government Printing Office.

U.S. Department of Transportation. Federal Railroad Administration. (2014). Mission statement. Retrieved from http://www.fra.dot.gov/Page/P0002

U.S. Department of Transportation. Federal Railroad Administration. (2014). Rail passenger / Amtrak Fatalities in 2010 [Data file]. Retrieved from http://safetydata.fra.dot.gov/OfficeofSafety/publicsite/on_the_fly_download.aspx

U.S. Department of Transportation. Surface Transportation Board. (2014). Definition of Class 1, 2 and 3 railroads. Retrieved from http://www.stb.dot.gov/stb/faqs.html

Van Horne Institute. (2004). *Calgary/Edmonton high speed rail: an integrated economic region.* Calgary, Alberta. The Van: Horne Institute.

Van Horne Institute. (2013). *Updated cost and ridership/revenue for Calgary/ Edmonton high speed rail.* Calgary, Alberta. The Van: Horne Institute.

Wisconsin Public Interest Research Group Foundation, (2010), *How a Faster Passenger Rail Network Could Speed Travel and Boost the Economy.*

World Bank. (2014, July 10). Retrieved from http://www.worldbank.org/en/news/press-release/2014/07/10/cost-of-high-speed-rail-in-china-one-third-lower-than-in-other-countries

World Market Intelligence News. (2014, July 15). *Ministry of transport - London to Toronto high-speed Rail line – Ontario.* Retrieved from http://ezproxy.lib.ucalgary.ca/login?url=http://search.proquest.com/docview/1552723667?accountid=9838

Yglesias, M. (2015). *Why American trains are so bad.* Retrieved from http://www.vox.com/2015/6/12/8764819/why-american-trains-are-bad

About the Contributors

Raj Selladurai is an Associate Professor of Management in the School of Business and Economics at Indiana University Northwest. He also serves as Small Business Institute (SBI) Director for the School of Business and Economics. His research focuses on the areas of servant leadership, leadership effectiveness, operations and supply chain management, high-speed rail and supply chain management, and business-healthcare collaboration. He has published several papers, chapters, and case studies in various management and operations management related business publications and journals including co-editing a book on "Servant Leadership: Research and Practice." He has over 27 years of academic and professional business experience including participating at the recent "Servant Leadership: Path to High Performance" Executive Education Seminar at the Darden School of Business, University of Virginia. He has also taught business management courses as Visiting Professor at universities in Austria and Switzerland. He has diverse leadership and administrative experience in various organizations including church, Bible Study Fellowship (BSF), university, community, business, and other professional service organizations.

Peggy Daniels Lee received her PhD from the George Washington University. She is a Clinical Assistant Professor of Operations and Supply Chain Management and Faculty Chair for Undergraduate Programs at the Indiana University Kelley School of Business, Indianapolis. Dr. Lee also holds a graduate certificate in Project Management and an Advanced Online Teaching Certificate from the Online Learning Consortium (formerly the Sloan Consortium). She teaches Operations Management and Sourcing at the undergraduate and graduate levels in face-to-face and online environments. Her industry experience includes a 20-year career in telecommunications and banking where she held multiple managerial positions, most notably as a project manager, leading national and international network infrastructure projects. Since 2007, Dr. Lee has done research and taught using 3-dimensional virtual worlds. Her work in this space has been presented at the Academy of Management, the Decision Sciences Institute, the Second Life Community Convention, the International Workshop on Enterprise, Organizational Modeling and Simulation (EOMAS), the Indiana University Center for Teaching and Learning, the virtual community Teach for Success, and many other organizations and venues. Her article in the Journal of Virtual Worlds Research is one of few that established the impact of virtual worlds on learning outcomes. Dr. Lee is also a section editor of the online Journal of Virtual Studies. Her academic work has been presented at the Decision Sciences Institute, the Academy of Management, the Council of Supply Chain Management Professionals, and the International Workshop on Enterprise and Organizational Modeling and Simulation (EOMAS). She has published in such outlets as the Journal of Business and Industrial Marketing, the Journal of Virtual Worlds Research, the Journal of Healthcare Management, and Supply Chain Forum, published by the University of Lausanne. She serves on the Board of Advisors for the Business Program at Ivy Tech, Indiana's Community College system, the Organizing Committee of EOMAS, serves the Midwest Decision Sciences Institute as treasurer and is he Education Chair for the Indianapolis Roundtable of the Council of Supply Chain Management Professionals. Dr. Lee is a former member of the board of the Delaware Valley Roundtable of the Council of Supply Chain Management Professionals and the board of the Northeast Decision Sciences Institute. In addition to her doctorate, Dr. Lee holds a BA in English Literature from the University of Michigan in Ann Arbor and an MBA from the Kenan-Flagler School of Business at the University of North Carolina at Chapel Hill.

George J. VandeWerken is Regional President of Providence Bank & Trust in the Chicago metro area. In addition to a lifelong career in community banking, George has also maintained an interest in public transportation, especially matters relating to rail service, and he currently serves as a member of the Board of Directors of the Indiana Passenger Rail Alliance.

***

Francis P. Banko received his BE in Electrical Engineering from SUNY Maritime College in 1988 and has 27 years of combined experience in the nuclear and transportation industries. He joined WSP | Parsons Brinckerhoff in 1997 after positions with CYGNA Energy Services, ICF Kaiser, and NJ Transit, the latter of which he served as the director of capital programs responsible for the procurement, overhaul, modification, and retrofit of NJ Transit's rolling stock fleet. In his capacity as a Vice President with WSP | Parsons Brinckerhoff, Frank manages the

Fleet and Facilities Division, and the delivery of professional services associated with the design, overhaul, and procurement of rolling stock. He currently serves as the Rolling Stock Manager for the California High-Speed Rail Project, where he is overseeing the procurement for 220 mph-capable trainsets. Frank is the company's 2010 William Barclay Parsons Fellow, for which he has co-authored the book titled Pioneering the Application of High-Speed Rail Express Trainsets in the United States (2013, Parsons Brinckerhoff Group, Inc.).

Jaydeep Balakrishnan, PhD, is a professor in the operations and supply chain management area at the Haskayne School of Business, where he has been since 1990. His research interests include the design of facilities to make them more efficient, improving the effectiveness of supply chain operations, and spreadsheet decision support modelling. He holds a PhD in operations management from Indiana University, an MBA from the University of Georgia and a BE in mechanical engineering from Nagpur University, India. He is also certified in production and inventory management by APICS. He has presented papers at various international conferences, has nearly 40 articles in reputable journals, and has co-authored an operations and supply chain management textbook. He has also provided seminars to professionals and academics internationally. Professor Balakrishnan has received several achievement awards and is listed in the Who's Who in Canadian Business published by the University of Toronto Press. He has been a visiting, teaching and research scholar in Hong Kong, Poland, Austria, Iran, and Singapore.

Rod Diridon, Sr. was the Emeritus Executive Director of the Mineta Transportation Institute from 1993 to 2014, Rod Diridon, Sr., was executive director of the Mineta Transportation Institute (MTI), and has chaired more than 100 international to local programs. Mr. Diridon was appointed in 2001 and 2005 by Governors Davis and Schwarzenegger to the California High Speed Rail Authority Board of which he is chair emeritus. He chaired the American Public Transportation Association's (APTA) High Speed Intercity Rail Committee and chairs the US High Speed Rail Association's (US HSRA) Board. In 1992, he served as the chair of APTA in Washington DC and for six years as North American vice chair of the International Transit Association (UITP) in Brussels. He provided over 50 sustainability keynotes, at US and international conferences. He received lifetime achievement awards from APTA, CUTC, National Association of Counties, San Jose State's College of Business, US HS Ground Transport Association, US HSRA, and others. He chaired nine successful rail project boards. The main train station was renamed the "San Jose Diridon Station" upon his retirement in 1995 after six terms as chair of both the Santa Clara County Board of Supervisors and Transit Board.

Dorothy E. Hagan received her Bachelor of Science in Public and Environmental Affairs from Indiana University in Fort Wayne, IN in 1988. She was hired by the City of Fort Wayne as a Neighborhood Planner in 1989. After 15 years in various positions with city governments, she returned to school to further her education. In 2011, she received her Doctorate in Law from Thomas M. Cooley University in Lansing, MI. She is once again employed in local government as a City Planner in Fort Wayne, Indiana. The Northern Indiana/Ohio Passenger Rail initiative is one of the current projects in which she is involved.

Richard Harnish is the Executive Director of the Midwest High Speed Rail Association (MHSRA), a member-supported non-profit organization advocating for fast, frequent and dependable trains linking the entire Midwest. Harnish co-founded MHSRA in 1993 and became its first executive director in February 2001 after leading a successful fundraising campaign. The Association is now the oldest and largest high-speed rail advocacy organization in the U.S.

W. Dennis Hodges is an active advocate for transportation and other community development projects. Born and raised in Gary, Indiana, W. Dennis Hodges lived in Chicago, Illinois for 20 years and Paris, France for three years. In 1975, he received a degree in English and history from Valparaiso University. Combined together, his education and residency set the stage for his professional life, which included doing market research for the multinational Club Med resorts. From these experiences, Dennis learned the profession of hosting and organizing corporate events for foreign diplomats. In his travels, Dennis has frequently ridden the high-speed passenger rail systems of Europe and found them to be an enormous convenience to the traveler. For this reason he organized, in 1992, the Indiana High Speed Rail Association, which has brought Indiana into a national debate around synergistic high-speed rail transportation issues. Dennis and the Association also advocate for the funding and implementation of the Midwest Regional Passenger Rail System, a nine-state, 3,000-mile passenger rail system. Dennis also advocates for community development through transportation and in this capacity has been a member of the Gary Chamber of Commerce board of directors, and its economic development and public policy committees. In addition, he routinely contributes opinions / columns to the Chicago Sun Times, The Times of Northwest Indiana and the Post Tribune of Northwest Indiana. Subsequently, Dennis is frequently called on to present opinions and to recommend policies on economic development and transportation projects.

Stephen C. Laffey is a Railroad Safety Specialist at the Illinois Commerce Commission in Springfield, Illinois. Mr. Laffey is the Chairman of the Transportation Research Board's Committee on Highway/Rail Grade Crossings (2009 – 2015). Prior to joining the ICC, Mr. Laffey worked at the Chicago Area Transportation Study (CATS) where he was Chief of Data Services. Since 2000, Mr. Laffey has been a member of the DuPage Railroad Safety Council. Mr. Laffey's primary research interests lie in transportation safety data and the use of Geographic Information Systems to address transportation safety issues. Mr. Laffey was awarded a B.S. in geology from Illinois State University (1986) and an M.S. in geography from Northern Illinois University (1994).

Jeff Morales is the Chief Executive Officer of the California High-Speed Rail Authority and has a distinguished record of experience managing large and complex transportation issues and projects. As the former Director of the California Department of Transportation, Morales managed a $10 billion program and more than 23,000 employees working to build, maintain and operate the largest state transportation system in the U.S. Morales most recently was Senior Vice President of Parsons Brinckerhoff, where he worked with transportation agencies across the United States and internationally to develop and implement major capital programs and is the past executive vice president of the Chicago Transit Authority where he spearheaded major reforms at the nation's second-largest transit agency. His experience at the federal level includes serving as a member of President-Elect Obama's transition team focusing on transportation, Vice President Al Gore's National Performance Review, the White House Commission on Aviation Safety and Security, the United States Department of Transportation and as U.S. Senate staff.

Eric C. Peterson is a policy advisor, currently addressing infrastructure funding and finance, public transportation, and intercity and high-speed passenger rail issues, and has played key roles in many major transportation improvement initiatives in the Washington, D.C. Metropolitan area over the past 30 years. He has held significant leadership roles on Capitol Hill, with national and regional transportation associations, and within the U.S. Department of Transportation where he was the first Deputy Administrator of the Research and Innovative Technology Administration. He led the effort to create the tax district that is financing the first phase of Washington Metro Rail's extension to Washington Dulles International Airport and Loudoun County, Virginia. He served for many years on the Arlington County, Virginia Economic Development Commission, and was a founding member of the

Washington Airports Authority's Advisory Board. An active member of the Transportation Research Board Intercity and High-Speed Rail Committee (AR010) and former member of the American Public Transportation Association's Intercity and High-Speed Passenger Rail Committee, he is also a member of the Board of Regents of the Eno Center for Transportation and on its P3 and NextGen working groups. He formerly served on the Eno Board of Advisors. He earned his undergraduate degree in Political Science from Nebraska Wesleyan University and his Masters degree from the School of Public Communication at Boston University.

Joseph Schwieterman, Ph.D., a professor of Public Service Management at DePaul University in Chicago, is a noted authority on transportation. He has twice testified on railroad issues before subcommittees of the U.S. Congress and is an officer of the Midwest High Speed Rail Association. His book, "When the Railroad Leaves Town: American Communities in the Age of Rail Line Abandonment," was a 2005 finalist of the George Hilton Book Award. Schwieterman is also regular guest who discusses transportation topics on First Business (syndicated public television) and National Public Radio. Schwieterman's mor recent book, "Terminal Town: An Illustrated Guide to Chicago's Airport, Bus Depots, Train Station and Steamship Landings," was released in 2014.

Peter Wallis is the President and CEO of the Van Horne Institute. His career has included senior positions at Canadian Airlines and in Ottawa as Legal Counsel for the Canadian Transport Commission and Executive Assistant to two federal ministers of transport. Mr. Wallis is a member of the Law Society of Upper Canada and serves on the Board of the Calgary International Film Festival and the Canadian Air Transport Security Authority. He is a Past Chairman of the Calgary Airport Authority, Past President of the Calgary Chamber of Commerce, and Past Vice-Chair of the Calgary Homeless Foundation. He serves as Honourary Consul for Mongolia in Alberta. Peter Wallis was recently awarded an honorary life membership by the Air Transport Association of Canada (ATAC). He was born in Toronto, Ontario, and holds a Bachelor of Arts from the University of Toronto, a Bachelor of Laws from Osgoode Hall and a Master of Laws from the University of London (England).

Jackson H. Xue received his BE in Mechanical Engineering from the Cooper Union for the Advancement of Science and Art in 2007, and was subsequently hired as a Catenary Engineer by WSP | Parsons Brinckerhoff. In 2009, he transferred to the company's Rail Vehicles Division, where he has worked on several high-speed rail initiatives and projects, lending rolling stock assistance domestically and internationally. He is currently involved in the California High-Speed Rail Project, with roles in trainset specification development and procurement support. Along with Frank, Jackson is a co-author of the book Pioneering the Application of High-Speed Rail Express Trainsets in the United States (2013, Parsons Brinckerhoff Group, Inc.).

Tianyuan Zhu is a PhD student of operation and supply chain management area at Haskayne school of Business, University of Calgary. She also holds an MSc in Transport from the Imperial College London and a BSc in Transportation Management from Dalian Maritime University, China. Tianyuan is currently supervised by Dr. Jaydeep Balakrishnan.

Index

N

O

P

Q

R

S

T

www.ingramcontent.com/pod-product-compliance
Lightning Source LLC
LaVergne TN
LVHW061542290926
849162LV00004B/876